# The
# New England
# BERRY BOOK

# The New England BERRY BOOK

## FIELD GUIDE & COOKBOOK

### Bob Krumm

YANKEE BOOKS
CAMDEN, MAINE

Cover and text design by Janet Patterson.

Printed in the United States of America.


Library of Congress Cataloging-in-Publication Data

Krumm, Bob, 1944–
    The New England berry book : field guide & cookbook / Bob Krumm.
       p.    cm.
    ISBN 0-89909-213-6 : $11.95
    1. Cookery (Berries)   2. Berries—New England—Identification.
I. Title.
TX813.B4K78   1990
641.6′47—dc20                                90-12241
                                                    CIP

# Contents

# Preface

Wild berries have been a part of my life for many years. I grew up in rural, southern Michigan, where blackberries, black raspberries, huckleberries, blueberries, elderberries, and currants grew in abundance. As a youngster I used to help my mother pick these wild crops and help with processing. I discovered that even though it was a lot of work to pick some of the wild berries, the effort was worth it. Mom cooked up some of the tastiest jams, jellies, pies, muffins, marmalades, and pancakes from the berries. A youngster would have lacked taste buds not to like all the berry goodies that she concocted.

As I reflect on my childhood, I realize that wild berries did more than provide tasty treats. They gave us an opportunity to do something together as a family. Even though at the time I then thought of berry picking as work, in retrospect I can remember that I took a lot of pride in filling my bucket with the juiciest, ripest blackberries or huckleberries.

Wild-berry-picking trips were an excuse for a picnic, and if we picked all that we needed in good time, we kids could go fishing in a nearby lake or stream. Wild berries have provided me with a host of pleasant memories of unhurried rural life and of a family enjoying the simple bounties of nature.

In 1987, I worked for L. L. Bean, Inc., in Freeport, Maine, as a fly-fishing instructor. While I was there, I happened to meet a new friend, Dot Heggie. We took quite a few walks together, and as we walked, I would point out the various berries that grew along the path.

As Dot and I strolled, I observed that New England and Michigan weren't that much different when it came to wild berries. I was flooded with memories of those luscious childhood treats. I explained to Dot the common uses for the berries, when they would ripen, and how to pick them. One day I announced that the flowering bush we were looking at was elderberry and it would yield lots of ripe berries in late September. When that season rolled around, I said, we would pick the clusters of ripe elderberries, go home, sit down and pull off each BB-sized berry and prepare for a feast. I told her that elderberry made some of the best imaginable jelly, pies and wine.

It was at this point that Dot said, "You ought to write a book." Of course I argued that I was too busy and that no one would read it, but in the end she prevailed. Thanks to her encouragement, the book exists.

Field guides to wild, edible plants can be found, and so can cookbooks that deal with natural fruits and vegetables, but a book that combines both a field guide with practical recipes is, to my knowledge, unique. I hope that you will find this book convenient and highly useful.

If you schedule camping trips around peak berry seasons, you can count on having plenty of good, wholesome snack and meal ingredients. You can plan on strawberries from late June into July, when the red and black raspberries will be ripening. Mid-July marks the start of the blueberry/huckle-berry season, which runs into August. In mid-August, the blackberries will be bountiful. By the time the blackberries are gone, the summer camping season is over, so you have made it through the season with the best of the berries.

I trust that *The New England Berry Book* will enable you to fully enjoy the many wild berries that grace New England. Once you pick a bucketful of them, try the recipes. I think they bring out the best flavors of the berries. Some jam and jelly recipes are old-fashioned; others are more modern and call for honey instead of sugar, and some are recipes for no-cook freezer jams or jellies. After you've concocted some of the recipes and tasted how luscious wild berries are, I know that you will have discovered why I treasure those childhood memories of wild berries and why I still forage for them today. I hope you will begin to collect your own pleasant memories of wild berries, too.

# Acknowledgments

I'd like to acknowledge a large number of persons for help with this book, but foremost are my parents, Donald and Emily Krumm. They taught me a great deal about wild plants and harvesting the natural bounty the good Lord created for everyone to enjoy.

I'd like to thank all those persons who submitted their favorite berry recipes. Janet Belanger of Buckfield, Maine, went out of her way to send us a bevy of berry recipes. Isabel Abbott of the Maine Blueberry Festival in Union was also a big help with recipes and information about the festival.

Dr. Sam Ristich, a retired Cornell University botanist, now living in Cumberland Center, Maine, was invaluable in so many ways. He contributed photos of berries I hadn't been able to photograph well, read the manuscript for botanical accuracy, provided professorial advice, added some anecdotes, helped to keep the manuscript taxonomically correct, and provided me with a cheery note of encouragement when I was down.

Fred Steele, a botanist from Tamworth, New Hampshire, read the manuscript and provided reams of helpful advice.

My proofreader, Pays Payson, of Falmouth, Maine, lent his expertise and encouragement.

Finally, my deepest thanks go to Dot Heggie of Falmouth. Without her help and gentle encouragement, I never would have written *The New England Berry Book*.

# Hazards

Picking wild berries can be a very enjoyable time, but there are some hazards of which you should be aware. There are insects that can pose problems. I mention some specific examples in the black currant chapter. Mosquitoes and black flies, the most common insect pests, can be warded off by applying a good insect repellent. Wild animals can be mighty protective of their berry patches. Bears like blueberries, huckleberries, and red raspberries. You're not likely to encounter a bear in southern New England, but if you're in the more remote sections of Maine, New Hampshire, and Vermont, keep an eye out for them. Remember, bears have the right of way. Let them have the patch and go look for another one.

Sometimes moose can be a bit cantankerous. Usually a cow moose is docile, but she can become obstreperous if you happen to get between her and her calf. Give a moose plenty of leeway. They don't particularly care for berries, so if you wait a bit, the moose will probably saunter away.

The biggest hazards are herbicides. Many private companies and governmental agencies use them to suppress the growth of broad-leafed plants. Most highway, power line, and railroad rights-of-way as well as recently harvested timber tracts will probably have been sprayed.

If the berry plants have off-color leaves—yellow or brown—or if the stems are brown, bent over, and curling, leave the fruit alone. In short, if the plants look like they are dying, *don't pick the berries*. Seek out an area where the plants are healthy looking. Herbicides can be carcinogenic, poisonous, or both.

Watch out for poison ivy and poison sumac, which have oils that can cause oozing blisters that itch like the dickens. Some people are not allergic to poison ivy or sumac, but most are. Learn what these plants look like, and try not to touch them. If you suspect that you have come into contact with them, wash the affected areas with strong soap as soon as possible.

There are some inedible berries, some require a lot of treatment before they are edible, and some are downright poisonous. Know your berries before you pick them! If you're not sure a particular berry is edible, don't pick it. In this book, I've tried to give you good verbal and photographic descriptions of these berries. Within each species there will be variation. The color of the berry may vary slightly and the size of the plant and size of the berry may vary from site to site—but the key identifying features will remain the same.

For instance, blueberries and huckleberries may be blue or black color. The bushes may be only one foot tall or may reach eight feet in height. But each berry will still have scalloped edges with five points at its base.

As the human population increases in New England, so does intolerance. Some people will not tolerate trespassing on their property for any reason; others will share their land if people ask permission. Before you enter private lands, ask permission. Most folks will be happy to let you pick berries on their land—especially if you offer to share some of what you find with them!

If you exercise a little caution and common sense, berry picking will become a source of enjoyment for you and your family. It will give you a chance to spend many pleasant days on family outings. The jams and jellies that you put up and the ones you freeze will remind you all winter of the wonderful berry-picking days of summer.

Enjoy the wild berries the good Lord created for us, but be aware that there are hazards to contend with in the berry patch.

# The
# New England
# BERRY BOOK

# Blueberries & Huckleberries

Blueberries and huckleberries are the stuff of folklore. They are so common throughout much of the northern United States that many folks have learned to relish them. I've picked blueberries and huckleberries in Maine, Michigan, North Carolina, Wyoming, and Montana. In New England, a good berry spot is almost a family treasure. You might lend someone your car, but never take anybody except family to your favorite blueberry spot! Beyond a doubt, the blueberry ranks as the favorite of New England berry pickers.

In Maine, blueberries are an industry, and the state has a blueberry festival each year in the town of Union. It is combined with an agricultural fair, and highlights include the coronation of Maine's Blueberry Queen, a blueberry pancake breakfast, and a blueberry pie-eating contest.

For me, blueberries represent the ultimate in trail cuisine. Fortunately, they ripen from early July through August—depending on the location and species—just when most people are on summer vacations. I can always count on blueberries to broaden my camp menu, or just to provide a trailside snack. Once I taught an outdoor school for young men that were from cities. When I talked about plants, they grew restive, although when I talked of fly fishing, they were right

1

with me. It was fortuitous that our first hike to a fishing spot went through a patch of blueberries. When I stopped to pick some for a snack, the young men looked at me apprehensively. I told them, "These are blueberries, give them a try."

In a minute I had six young blueberry addicts. From then on, our hikes looked more like skirmish lines as they fanned out from the trail. Invariably one would holler "Blueberries!" Everyone would scramble to the site and proceed to pick the place clean. Those once uninterested young fellows even made blueberry jam! I wonder what their parents thought about my outdoor school when their sons came tromping in, proudly displaying jars of blueberry jam they had made.

Blueberries are easy to identify and widely available. On a weekend fishing or hiking trip, take time out to enjoy a patch of blueberries. My twin sons, Clint and James, used to complain about blueberry picking—until we collected a gallon of berries for jam and pies—then they thought that blueberries were the best food since buttered popcorn. My sons are now young men and still look forward to berry-picking expeditions with their dad. From mid-July through mid-August, I'd say your chances were very high of finding a quart or two of berries in most parts of New England. About the only thing that can eliminate a blueberry or huckleberry crop is a severe late spring cold snap. All the birds, bears, and other berry-eating animals in New England can't eat enough to deplete the supply!

Blueberries belong to the genus *Vaccinium* and its closely related cousin, huckleberry, belongs to the genus *Gaylussacia*. Both belong to the same family *Ericaceae*. The basic difference between the huckleberry and blueberry is the seeds: blueberries have many small seeds; huckleberries have ten larger seeds. While there are other minor differences between the genera, the berries look pretty much the same, and taste the same, so trying to describe the differences between the genera would only lead to confusion. The bottom line is that they taste great! Consequently, blueberries and huckleberries are lumped together in this book, and I will use the word blueberry to refer to them both.

New England has several species of blueberries ranging from lowbush to highbush. Some species reach a height of five feet, others attain a ground-hugging height of three to six inches. All species prefer open to subdued sunlight; none thrives in dense shade. Blueberries prefer acid soils that range from moist to soggy.

Blueberries blossom in May and mature from early July to late August. Of course, those blueberry species growing in southern New England mature earlier than those in northern or higher-altitude areas. There might be as much as a two-week difference in maturity between southern and northern regions for a given species.

Most blueberries are a powdery blue when ripe but some turn black. The fruits range in size from slightly smaller than a pea to half again as large as a pea. All blueberries have a "navel"—a scalloped edge that looks slightly like a scalloped pie crust. This navel has five points—the remains of the flower part called the calyx. The calyx is positioned above the ovary. As the berry grows, the calyx is left on the "top" of the berry. You'll see the calyx in the same place on apples, pears, june-berries, and crab apples.

Blueberries have a small, slightly pointed, pale-green leaf. The leaf is slightly leathery with margins (edges) varying from entire (no serrations) to finely serrated (small teeth). The leaves are arranged alternately.

All blueberries are perennials, lasting for many years. Carbon dating of some blueberry clones in New Jersey has shown them to be at least two thousand years old and still growing! They have woody branches and stems. Blueberries do not have thorns.

While there are some plants that vaguely resemble blueberries, none has the combination of characteristics—the scalloped base, the light green leaves, the light blue to black color of the berries, the lack of thorns, and the woody twigs—

3

so if you look for these characteristics, you won't pick the wrong fruits.

Blueberries keep well in the refrigerator, but do not wash them before storing. Keep them covered loosely and they'll last up to two weeks. They also freeze well. Spread them on trays or cookie sheets, put the trays in the freezer until the berries are frozen, then put the berries in quart-sized freezer bags and return them to the freezer. By freezing them this way, you won't have to contend with a frozen block of blueberries.

According to the Maine Department of Agriculture, you can use frozen blueberries in recipes exactly like fresh ones. They emphasize, however, *do not wash the berries before freezing them and do not defrost them unless the recipe specifies it.*

With plenty of frozen blueberries you can try out the numerous blueberry recipes throughout the year, instead of trying to make them all during the harvesting season.

# Recipes

Blueberries have a wide variety of uses, but they are hard to beat just as is. They are simply scrumptious eaten raw, with a little milk and sugar, sprinkled over cold cereal, in fruit salads, and as toppings for waffles and pancakes.

Then, too, blueberries are great cooked—jams, jellies, syrups, pies, tarts, pancakes, muffins, compotes, and turnovers, to name a few.

## WILD BLUEBERRY PUMPKIN MUFFINS

1$2/3$ cups flour
1 teaspoon baking soda
1/2 teaspoon baking powder
1 teaspoon cinnamon
1/2 teaspoon allspice
1 cup canned pumpkin
1/2 cup evaporated milk

1/3 cup shortening
1 cup firmly packed light
  brown sugar
1 egg
1 cup wild blueberries
1 tablespoon flour

*Streusel topping*
2 tablespoons flour
1 tablespoon sugar

1/4 teaspoon cinnamon
1 tablespoon butter

Preheat oven to 350 degrees. Combine flour, baking soda, baking powder, cinnamon, and allspice. Combine pumpkin and evaporated milk until blended. Cream shortening and sugar in large mixer bowl. Add egg. Beat until mixture is fluffy. Add flour alternately with pumpkin mixture, beating well after each addition. Combine wild blueberries with flour. Gently stir them into the batter. Fill 18 paper-lined muffin tins three-quarters full. Sprinkle streusel over top of muffins and bake for 40 minutes (until toothpick comes out of center clean).
Yield: 18 small muffins or 12 large ones
*Helen Fitzgerald, Lansing, Michigan*

## BLUEBERRY GRUNT PUDDING

2 cups blueberries
1 cup sugar
1 teaspoon lemon juice
1/4 stick margarine or butter, melted
1 cup flour

1 1/2 teaspoons baking powder
1/4 teaspoon salt
1 egg
1/2 cup brown sugar
1/2 cup hot milk

Preheat oven to 400 degrees. Combine blueberries, sugar, lemon juice, and melted margarine in bottom of baking dish.

In a mixing bowl combine flour, baking powder, and salt. Blend egg, brown sugar, and hot milk. Combine the wet and dry ingredients and place the batter over the top of the blueberry mixture.

Bake for 30 minutes.

*Adapted from Marion Averill Lewis,* The Norway Pines House Cookbook

## BLUEBERRY CAKE

2 eggs, beaten lightly
1/2 cup molasses
4 tablespoons sour milk
1 teaspoon salt
2 cups flour

1 cup sugar
1/2 cup melted butter
1 teaspoon baking soda (dissolved in milk)
2 cups blueberries

Blend eggs, sugar, melted butter, and sour milk (with the dissolved baking soda). Dust the blueberries with two tablespoons of flour. Blend remaining flour into wet ingredients and fold in blueberries. Grease and flour a 9 x 13 baking pan. Pour batter into it and bake in a slow oven (300 degrees) for 1 hour. After the cake has been in the oven a few minutes, sprinkle a little cinnamon and sugar on top.

*Adapted from Marion Averill Lewis,* The Norway Pines House Cookbook

## BLUEBERRY JAM

| | |
|---|---|
| 2 quarts blueberries | 6 cups sugar |
| 1/4 cup lemon juice | 1 box pectin |

Crush blueberries. Measure 3$^3$/4 cups crushed berries and place in 6- or 8-quart saucepan or kettle. Add lemon juice and pectin to berries and stir thoroughly. Bring mixture to rolling boil over high heat, stirring constantly. Add sugar. While stirring constantly, return to rolling boil. Boil hard for 1 minute, again stirring constantly to prevent scorching.

Remove from heat; skim off foam. After skimming, pour or ladle the jam into sterilized 8- or 12-ounce jars. Wipe off jar rims and threads. Attach two-piece metal lids. Tighten securely. Submerge in boiling water bath for at least 5 minutes.

## BLUEBERRY-APPLE COMPOTE

| | |
|---|---|
| 6 cups peeled, sliced apples | 1 tablespoon grated orange |
| 3 cups apple juice | peel |
| 1 tablespoon lemon juice | 1 tablespoon cinnamon or |
| 2 cups sugar | 6 tablespoons Grand |
| 6 cups fresh or frozen | Marnier |
| blueberries | |

Combine apples, apple juice, lemon juice, and sugar in a saucepan. Cook over low heat about 10 minutes, until apples soften and sugar is completely dissolved. Remove from heat and fold in blueberries and orange peel. Add cinnamon or Grand Marnier.

Ladle hot fruit into clean, hot pint jars, leaving 1/2-inch headroom. Seal. Process in boiling-water canner 10 minutes. Let cool, undisturbed, for 12 hours. Check seals. Store in cool, dry place.

Yield: approximately 6 pints

*Andrea Chesman.* Summer in a Jar: Making Pickles, Jams & More

## CAPE COD BLUEBERRY GRUNT

2 cups blueberries
1/4 cup water
2 tablespoons sugar
3/4 cup all-purpose flour

1/2 teaspoon salt
2 1/2 teaspoons baking powder
3/4 cup milk

Simmer blueberries in water until the berries soften. Place cooked berries with sugar in a deep baking dish. In a separate bowl prepare biscuit dough by combining flour, salt, baking powder, and milk. Cover the blueberries with the biscuit dough. Place baking dish in an eight-quart or larger pot. Add boiling water to pot so that the water level is about one inch from the top of the baking dish. (Don't put the water *in* the baking dish.) Cover the pot and boil for 1 hour. Add water as needed. Serve with cream or whipped cream.

*Dot Heggie, Falmouth, Maine*

## YOGURT HUCKLEBERRY PIE*

2 8-ounce containers vanilla
  yogurt
1 cup huckleberries (or
  blueberries)

1 8-ounce container Cool-
  Whip
8- or 9-inch baked pie shell

Fold fruit into yogurt, then fold in Cool-Whip. Spoon mixture into crust. Freeze until firm (about 3 hours). Remove from freezer about 30 minutes before serving. Store leftover pie in freezer.

Huckleberry Recipes. *Compiled by Swan Lake Women's Club, Swan Lake, Montana*

*Can substitute blackberries, black raspberries, juneberries, wild strawberries.

## CAMPFIRE HUCKLEBERRY PIE

1 rainhat full of plump berries!
1 package instant vanilla
  pudding

Pie crust (whole-wheat flour,
  a little Bisquick, oil, honey,
  water, and powdered milk)

Mix crust and bake in reflector oven or with hot rocks. Fill with vanilla pudding and berries.
Huckleberry Recipes. *Compiled by the Swan Lake Women's Club, Swan Lake, Montana*

## CRAZY CRUST BLUEBERRY PIE*

*Pastry*
1 cup all-purpose flour
2 tablespoons sugar
1 teaspoon baking powder
1/2 teaspoon salt

3/4 cup water
2/3 cup shortening
1 egg

*Filling*
1 quart blueberries
1 cup sugar

1 teaspoon cinnamon
1/4 teaspoon nutmeg

Preheat oven to 425 degrees. In small mixer bowl, combine flour, sugar, baking powder, salt, water, shortening, and egg. Blend well at lowest speed; beat 2 minutes at medium. Spread batter in 10-inch or 9-inch deep-dish pie pan.

Prepare batter by combining blueberries with sugar and spices. Carefully spoon filling into center of batter. Do not stir. Bake for 40 to 45 minutes until crust is golden brown.
*Emily Krumm's modification of a Pillsbury Flour recipe.*

*This recipe would also work well with blackberries, elderberries, black raspberries, or juneberries.

## BLUEBERRY CRISP

4 cups fresh or frozen blueberries
2 to 4 tablespoons sugar
2 teaspoons lemon juice
1/4 cup butter or margarine
1/3 cup packed brown sugar

1/3 cup all-purpose flour
1/4 teaspoon cinnamon
dash of salt
3/4 cup quick oats or 1/4 cup
   old-fashioned rolled oats

Preheat oven to 375 degrees. Place blueberries in buttered 8-inch-square baking dish; sprinkle with 2 to 4 tablespoons sugar and lemon juice.

In a medium bowl mix together butter or margarine, brown sugar, flour, cinnamon, and salt till mixture is crumbly. Stir in rolled oats, sprinkle mixture evenly over blueberries. Bake for 35 to 40 minutes or till topping is nicely browned. Serve warm with whipped cream.

Yield: 6 to 8 servings

*Maine Department of Agriculture,* Fresh Maine Blueberries Go Wild In Your Kitchen

## EXCEPTIONAL BLUEBERRY CAKE

*Cake*
1 1/2 cups flour
1/2 cup butter, softened
1 egg
1/2 teaspoon nutmeg
1/2 cup sugar

1/2 teaspoon baking powder
1 teaspoon vanilla
1 quart fresh or frozen
   blueberries

*Topping*
2 cups sour cream
1/2 cup sugar

2 egg yolks
1 teaspoon vanilla

Preheat oven to 350 degrees. Mix together all but the blueberries. Pour into a well-buttered 9-inch springform pan. Sprinkle blueberries over the top. Blend the topping ingredients well. Pour over batter. Bake 60 minutes or until edges are lightly browned.

*Heidi Hitchcock, Gorham, Maine*

## ANNE'S BLUEBERRY SCONES

2 cups all-purpose flour
1/2 cup sugar
2 teaspoons baking powder
1/2 teaspoon baking soda
1/2 teaspoon salt
1/2 cup butter or margarine

2 eggs, beaten
1/2 cup buttermilk or
   soured milk
1 cup fresh or frozen
   blueberries

Preheat oven to 375 degrees. Sift together flour, sugar, baking powder, baking soda, and salt into a large mixing bowl. With pastry blender or two knives used scissors-fashion, cut in butter or margarine till mixture resembles fine crumbs. Add beaten eggs and enough buttermilk or soured milk to make a soft dough. Carefully fold in blueberries. Spread dough in greased 9 x 9 baking pan. Bake in oven for 25 to 30 minutes or till no batter remains on an inserted wooden pick.

Serve hot, or split and toast when cold.

Yield: 12 servings

*Maine Department of Agriculture,* Fresh Maine Blueberries Go Wild In Your Kitchen

## BLUEBERRY DUFF

1/3 cup brown sugar
1/3 cup molasses
1/3 cup butter
2 cups flour
1 teaspoon baking powder

1/2 teaspoon soda
1/2 teaspoon salt
1/3 cup milk
11/2 cups blueberries

Blend brown sugar, molasses, and butter. Mix in remaining ingredients except blueberries. Butter mold (a 2-pound coffee can or shortening can will do) and layer batter and berries until two-thirds full. Cover and steam on a trivet in kettle of boiling water 11/2 hours. Serve hot with foamy sauce, hard sauce, or ice cream.

*Isabel Abbott, Maine Blueberry Festival, Union, Maine*

## BLUEBERRY COFFEE CAKE

*Cake*

3/4 cup sugar
1 egg
1/4 cup vegetable oil
1/2 cup milk
2 cups plus 2 tablespoons
   sifted flour

2 teaspoons baking powder
1/2 teaspoon salt
2 tablespoons lemon juice
2 cups blueberries (fresh or
   frozen)

*Crumb topping*

1/2 cup sugar
1/3 cup flour

1/2 teaspoon cinnamon
1/4 cup soft margarine

Preheat oven to 375 degrees. Combine sugar, egg, and oil and mix well. Sift together flour, baking powder, and salt and add, alternately with milk, to sugar mixture. Add lemon juice and blueberries. (Gently fold in blueberries.) Spread batter in greased and floured 9 x 9 pan. Sprinkle with crumb mixture. Bake 45 to 50 minutes or until lightly browned and tests done. This coffee cake freezes well, so it can be made ahead of an occasion and saved.

*Pat Joyce, Wilton, Maine*


## BLUEBERRY DESSERT

*Berries*

2 cups blueberries or huckleberries, fresh or frozen

1 to 2 tablespoons lemon juice
1 teaspoon cinnamon

Butter 8 x 8 baking pan. Put blueberries into pan, dribble lemon juice over the top, and sprinkle cinnamon over berries.

*Cake*

3/4 cup sugar
3 tablespoons butter or margarine
1 cup flour

1 teaspoon baking powder
1/4 teaspoon salt
1/2 cup milk

Preheat oven to 375 degrees. Blend sugar with butter or margarine. Sift together dry ingredients. Add dry ingredients and milk alternately to sugar and butter mixture. Spread batter over the blueberries.

*Topping*

| | |
|---|---|
| 1 cup sugar | 1¹/₂ tablespoons cornstarch |
| dash of salt | 1 cup boiling water |

Mix sugar, salt, and cornstarch thoroughly and sprinkle evenly over batter. Pour boiling water over top. Bake 1 hour. This recipe goes great with ice cream or whipped cream. It tastes best when served hot.

## BLUEBERRY GINGERBREAD

| | |
|---|---|
| ¹/₂ cup shortening | 1 cup soured milk or |
| ¹/₄ teaspoon salt | buttermilk |
| 1 cup sugar | 3 tablespoons molasses |
| 1 egg | 1 cup blueberries or |
| ¹/₂ teaspoon ginger | huckleberries |
| 1 teaspoon cinnamon | 3 tablespoons sugar plus |
| 2 cups flour | ¹/₄ teaspoon cinnamon—for |
| 1 teaspoon baking soda | topping |

Preheat oven to 350 degrees. Cream together shortening, salt, and sugar. Add unbeaten egg and blend until light and creamy. Sift ginger, cinnamon, and flour together.

Measure baking soda into buttermilk or soured milk. (If you need to make soured milk, add 2 tablespoons vinegar to 1 cup whole milk.) Stir to dissolve. Add sifted dry ingredients and soured milk alternately to creamed mixture. Add molasses. Add blueberries. Turn batter into a greased and floured 9- x 9-inch pan. Sprinkle sugar and cinnamon mixture over top of batter. Bake 50 to 60 minutes.

## NEW ENGLAND BLUEBERRY PIE

Pastry for 2-crust pie (see juneberry recipes)

1 cup sugar
3 tablespoons cornstarch
4 cups blueberries
dash of salt

$1/4$ teaspoon nutmeg
$1/4$ teaspoon cinnamon
1 tablespoon butter

Line pie plate with pastry. Preheat oven to 425 degrees. Mix together sugar and cornstarch. Spread about one-fourth of mixture on lower crust. Fill with blueberries. Sprinkle remainder of sugar mixture over them. Add salt. Sprinkle with nutmeg and cinnamon. Dot with butter. Place top crust on pie, flute edges, and cut slits. Bake at 425 degrees for 15 minutes, reduce heat to 350 degrees, and bake 30 to 40 minutes more.

## BLUEBERRY CONSERVE

2 oranges
juice of 1 lemon
$1/2$ cup water

1 quart blueberries
sugar

Peel the oranges. Cut rind into thin strips. Remove white inner skin and cut fruit into thin slices. Add lemon juice, water, and blueberries. Bring to a boil, then measure. Add to measured fruit mixture in kettle two-thirds as much sugar as fruit. Simmer until mixture is thick. Pour into sterilized jars, wipe rims and threads clean, securely attach two-piece metal lids and place in boiling water bath for 5 minutes.

*Adapted from Clarissa M. Silitch, ed.* The Forgotten Arts: Making Old-Fashioned Jellies, Jams, Preserves, Conserves, Marmalades, Butters, Honeys & Leathers

## HUCKLEBERRY BARS

1/2 cup margarine
11/2 cups sugar
2 eggs
1 teaspoon vanilla
2 cups flour

2 teaspoons baking powder
1/4 teaspoon salt
1 cup chopped nuts
2 cups huckleberries or
    blueberries

Preheat oven to 350 degrees. Melt margarine, remove from stove, and stir in sugar. Cool slightly. Add eggs and vanilla. Beat well. Combine dry ingredients. Blend into egg mixture. Add the nuts and stir in berries. Spread batter in a greased 9 x 13 pan. Bake for 35 minutes. Cut into bars.
Huckleberry Recipes. *Compiled by Swan Lake Women's Club, Swan Lake, Montana*

## PICKLED SPICED BLUEBERRIES

5 pounds blueberries
1 cup vinegar
1 cup water
6 cups sugar

1 tablespoon whole allspice
1 tablespoon whole cloves
1 tablespoon stick cinnamon
    pieces

Wash berries and drain. Combine vinegar, water, and sugar. Tie spices loosely in cheesecloth bag and boil in vinegar mixture 5 minutes. Add well-drained berries and simmer 5 to 10 minutes. If cooked too long, berries will have a shriveled appearance. Remove spices. Pour into sterilized jars and seal. Immerse in boiling water bath 15 minutes.
Yield: about 8 pints
*Dot Heggie, Falmouth, Maine*

## BEST BLUEBERRY CUPCAKES

$1/2$ cup margarine
$3/4$ cup sugar
2 eggs
2 teaspoons baking powder
$1/2$ cup milk

2 cups flour
$1/2$ teaspoon vanilla
$1^{1}/_{2}$ cups blueberries
1 tablespoon sugar mixed
    with $1/2$ teaspoon cinnamon

Preheat oven to 375 degrees. On low speed, cream margarine and sugar until fluffy. Add eggs, beat well. Add dry ingredients alternately with milk and vanilla. Mash $1/2$ cup berries and stir in by hand. Add remaining berries whole and stir by hand. Grease muffin pans with Pam and spray top of pan also. Pile batter even with the top of each muffin cup. Mix sugar and cinnamon and sprinkle a little on each cupcake. Bake about 30 minutes.
*Pat Champagne, Biddeford, Maine*


## BLUEBERRY SYRUP

$1/2$ cup brown sugar
1 tablespoon cornstarch
dash of salt
$1/2$ cup water

2 cups blueberries
1 tablespoon lemon juice
1 teaspoon grated lemon rind

Combine dry ingredients. Stir in water. Add blueberries and bring to boil. Simmer until clear and thickened, about 5 minutes, stirring continuously. Remove from heat, stir in lemon juice and rind.

This syrup is delicious on pancakes, waffles, ice cream, and French toast.

## BLUEBERRY DUMPLINGS

2 quarts blueberries
1/2 cup water

*Dumplings*

2 cups flour
4 teaspoons baking powder
1/2 teaspoon salt

3/4 cup milk and water combined

Stew blueberries with water.

Combine dry ingredients and then mix in milk and water. Drop dumplings on top. Cover, and steam for 12 minutes. Cooking to Beat the Band. *Compiled by Band Mothers Club, Deering High School, Portland, Maine*

## NEW HAMPSHIRE BLUEBERRY CAKE

2 eggs, separated
1 cup sugar
1/2 cup shortening
dash of salt
1 teaspoon vanilla

1/3 cup milk
1 1/2 cups sifted flour
1 1/2 teaspoons baking powder
1 1/2 cups fresh blueberries

Preheat oven to 350 degrees. Beat egg whites until stiff. Add some sugar to keep the egg whites stiff. Cream together shortening. Add salt and vanilla. Add remaining sugar gradually. Add unbeaten egg yolks. Beat until thoroughly blended and creamy. Alternately add milk and dry ingredients, blend together well. Coat blueberries with a tablespoon or so of flour. Fold blueberries into batter. Pour into a greased 8- x 8-inch pan. Sprinkle top of batter lightly with granulated sugar and cinnamon. Bake 50 to 60 minutes.

## BLUEBERRY MUFFINS

| | |
|---|---|
| ¹/₄ cup shortening | 1 egg |
| 1¹/₂ cups flour | ¹/₂ cup milk |
| ¹/₂ cup sugar | 1 cup blueberries, washed and |
| ¹/₂ teaspoon salt | drained |
| 3 teaspoons baking powder | |

Preheat oven to 400 degrees. Grease muffin cups. Melt shortening. Sift together flour, sugar, salt, and baking powder. Beat egg and milk. Stir in dry ingredients. Blend in shortening. Add berries last. Fill the muffin cups two-thirds full and bake 20 to 25 minutes.
Yield: 12 muffins
*Emily Krumm, Eaton Rapids, Michigan*


## MAINE BLUEBERRY MUFFINS

| | |
|---|---|
| 2 cups flour | 1 cup milk |
| ¹/₄ cup sugar | ¹/₄ cup salad oil |
| ¹/₂ teaspoon salt | 2 teaspoons lemon juice |
| 3 teaspoons baking powder | ³/₄ cup blueberries |
| 1 egg | |

Preheat oven to 425 degrees. Sift together dry ingredients. Beat egg well. Add milk, oil, and lemon juice. Combine wet and dry ingredients quickly. Mixture should be lumpy. Fold in blueberries, gently. Fill greased muffin tins two-thirds full. Bake about 20 to 25 minutes.
*Maine Blueberry Festival, Union, Maine*

## BLUEBERRY WINE

8 quarts blueberries
2 gallons boiling water
1 tablespoon granulated wine
   yeast, dissolved in 1/2 cup
   warm water

juice of 1 lemon, freshly
   squeezed
16 cups white cane sugar

1. Pick over blueberries; remove any stems, leaves, and blemished berries.
2. Pour boiling water over berries. Cover and allow to stand 24 hours, stirring often.
3. Place a large colander over a large bowl and cover with two thicknesses of finely woven cheesecloth. Strain juice. Rinse out crock and wipe clean. Return strained juice to crock.
4. Add sugar, dissolved wine yeast, and lemon juice. Cover well with a large lid (or cover with a polyethylene sheet and tie it down tightly around rim). Allow to rest in a warm place (65 to 70 degrees) for 3 days.
5. Pour into an opaque fermentation jug or wrap a large, clear glass jug with brown paper to seal out light so as to protect color.
6. Fit with an air lock and allow to ferment until bubbling has stopped. Siphon out into sterilized bottles; cork, and store. You may wish to decant before serving at room temperature.

Yield: about 10 quarts

*Jean Ann Pollard.* The New Maine Cooking: Serving Up the Good Life

## DOWN EAST BLUEBERRY TORTE

| | |
|---|---|
| 4 egg whites | 1/2 teaspoon vanilla |
| 1/4 teaspoon cream of tartar | 2 cups fresh blueberries |
| 1 cup sugar | whipped cream or yogurt |

Preheat oven to 275 degrees. In medium bowl beat egg whites until frothy; add cream of tartar and beat until stiff but not dry. Gradually add sugar, a spoonful at a time, beating until smooth and glossy. Beat in vanilla. Spread meringue on bottom and sides of a buttered 9-inch pie plate, hollowing out center and being careful not to spread too close to the rim. Bake in oven for 1 hour or until firm and dry to the touch but not browned. Turn off heat and cool in oven with the door open. Refrigerate until serving time.

Cover meringue with blueberries; sprinkle lightly with sugar if desired. Garnish with whipped cream or yogurt.
Yield: 8 to 10 servings
*Maine Department of Agriculture,* Fresh Maine Blueberries Go Wild In Your Kitchen

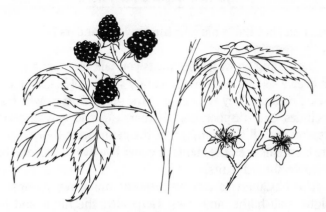

# Blackberries

People hold different opinions about blackberries. Some folks relish them, cultivate and nurture them; others try to eradicate them. Some seek out blackberries; others avoid them. Blackberries are a noxious weed in New Zealand. I find it amusing because there are a number of New Zealanders who harvest the escaped North American import. In 1984, I spent a month in New Zealand visiting Tom and Pi Kemper with whom I had been good friends in Jackson Hole, Wyoming. In 1981, the Kempers emigrated to New Zealand to start a whitewater float trip company.

One day, when we were returning from a fishing trip near Lake Taupo, Tom stopped his vehicle and announced we were going to go blackberry picking. The three of us grabbed some plastic buckets and headed toward a big briar patch where we picked three or four quarts of some of the most beautiful blackberries I had ever seen. When we went back to the Kemper home, I got busy and prepared a pie crust. Some of Pi's female friends found it amazing that a man would make pies. Even more astounding to them was how good the "noxious weed" pie was. Most of the ladies had no idea that blackberries were good to eat. If I recall right, the Kempers and I had but half a pie left; the New Zealanders talked us out of the remaining pie!

Depending on which botanists you talk to, there are as few as twelve species or as many as fifty species of blackberries in New England. They all belong to the genus *Rubus*, which also includes the raspberries, dewberries, and boysenberries. Essentially, all the members of the genus have thorny stems, are biennials (i.e., the stems or canes live two years), and have thimble-shaped fruits.

The blackberries can be lumped into three groups: the highs, half-highs, and lows. Generally, the highs and lows have good fruit while the half-highs usually have a sparse crop. This book concentrates on the high grouping and briefly discusses the lows.

Blackberries blossom in June and the berries mature in August. As it ripens, the fruit turns from green to red to a glossy black. When picked, a blackberry retains the receptacle (i.e., the core of the fruit). Raspberries do not retain their receptacles so the white "core" remains on the stem when they are picked.

Blackberries have stout, ribbed, red-hued canes (stems) with triangular thorns that are quite substantial. The number of thorns on the canes ranges from a few sparse ones to several per inch. Some canes are light green; some tend to light brown, but most have streaks of red to solid red color.

The leaves are palmately compound (that is, the leaflet stems originate from the same point) with three to five leaflets. The leaves are arranged alternately.

Blackberries prefer deep, well-drained soils, but make do in sandy soils, clay soils, and in waste areas. They grow at the edge of woods, along rights-of-way, in forest openings, or in any other place where there is full to moderate amounts of sunlight. Like most invading plant species, blackberries die out in the full shade of a mature forest.

The biggest problem with picking blackberries is navigating through those stout, thorny branches. Many berry pickers wear brush pants, some use heavy leather gloves to pull the branches aside, some use leather forearm guards, some even use hedge shears to clear a path. The hedge shears are a bit extreme, but some people will do anything for a bucket of berries.

Closely related to blackberries are dewberries. The fruits look much the same, and the only readily recognizable difference is that dewberries are recumbent; that is, they grow close to the ground and thus belong to the "low" group.

This growth form limits the sun-loving berry bushes to open fields and rights-of-way. Dewberries ripen in early August and have black, thimble-shaped fruits with a luscious blackberry taste.

In most years, it's relatively easy to pick two to four quarts of blackberries in two to three hours' time. The trick is to find a good patch—if you do, your chances are nearly a hundred percent that you'll fill a one-gallon berry bucket.

To find the perfect patch takes some preseason scouting. Look for thick, red-hued canes that are loaded with blossoms in late May or early June. Check back in July to see if the berries have set and that the patch hasn't been sprayed. Try to have a half dozen or so "perfect patches" lined up so that you can pick each at the peak of ripeness. Those patches that are more sunlit will ripen first; the shaded ones mature later in August. By following this type of a schedule, you can pick succulent, savory blackberries in abundance for nearly a month.

When you get a bountiful picking of blackberries, you can either make them into jam, jelly, and other fresh fruit recipes or freeze them—blackberries will not keep long. They do freeze easily. All you have to do is wash and drain them, then place them on a tray, put it in the freezer, and let the berries freeze individually. Then place the frozen berries in a plastic freezer bag.

You can also pack them with three-fourths cup of sugar per quart of berries or freeze them in a sugar syrup. How you preserve them depends on how you intend to use them.

Dry pack lends to jams, jellies, sauces, and flavorings, while the sugar or syrup pack is better suited for desserts and compotes.

Whichever way you preserve blackberries, you can count on having flavorful berries ready for cooking throughout the winter. There's nothing like having fresh blackberry cobbler in January!

# Recipes

## BLACKBERRY FROZEN YOGURT*

| | |
|---|---|
| 7 cups fresh blackberries | 1/2 cup sugar |
| 1/4 cup cold water | 1 16-ounce container vanilla |
| 1 envelope unflavored gelatin | low-fat yogurt |

Process blackberries in food processor until mixture is smooth—set purée aside.

Sprinkle gelatin over cold water, let stand 5 minutes. Place gelatin over low heat and cook 4 minutes or until gelatin is dissolved, stirring constantly. Remove from heat, add yogurt and sugar. Blend well. Cover and chill, then add berry mixture. Pour blackberry-yogurt mixture into freezer of ice cream maker. Follow manufacturer's instructions to prepare.
Yield: 12 servings
*Charlene Dihel, Longmont, Colorado*

*Raspberries or strawberries can be substituted. If strawberries are used, add 1 teaspoon vanilla to mixture before freezing.

## BLACKBERRY FLUMMERY

| | |
|---|---|
| 1 quart blackberries | dash of cinnamon |
| 1/2 cup hot water | 3 tablespoons water |
| 1 cup sugar | 2 tablespoons cornstarch |
| dash of salt | |

Combine blackberries, hot water, sugar, salt, and cinnamon in a saucepan and bring to boil. Simmer gently 5 to 8 minutes. Add 3 tablespoons water to cornstarch, and blend into berry mixture. Cook until translucent (3 to 5 minutes). Strain through food mill and cool. Serve cold with cream.
Yield: 5 to 6 servings
*Mrs. A. Holmes Stockly, Falmouth, Maine*

## BLACKBERRY CLAFOUTI

*Crust*

1¹/₂ cups all-purpose flour

1 stick unsalted butter, chilled,
  cut in pieces

¹/₄ teaspoon salt

¹/₄ cup ice water

*Filling*

Finely crushed vanilla wafers

2 cups blackberries

*Topping*

4 large eggs

8 ounces sugar (1 cup plus
  8 teaspoons)

³/₄ cup melted sweet butter,
  heated until lightly browned

¹/₃ cup all-purpose flour

1 teaspoon vanilla

crème fraîche

Position rack in lower third of oven and preheat to 350 degrees. Generously oil 11-inch tart pan with removable bottom.

Prepare crust: Combine flour, butter, and salt. Blend with electric mixer until crumbly and size of peas. Add water and continue to mix well until mixture comes clean from side of bowl and forms a ball. Flatten into 8-inch circle, wrap in plastic, chill 30 minutes. Place pastry on well-floured surface and roll into 12-inch circle. Starting at edge of circle, roll pastry around rolling pin and gently transfer to prepared pan. Fit into pan, trim excess. Sprinkle bottom of crust with wafers, and fill with berries.

Prepare topping: Combine eggs, sugar, butter, flour, and vanilla. Beat until smooth. Pour over berries, bake in lower third of oven until top is golden and crust begins to shrink away from sides of pan, about 60 to 65 minutes. If top browns too quickly, cover loosely with foil. Serve warm with crème fraîche.

Yield: serves 8

*Helen Abbott, Moultonboro, New Hampshire*

## NO-COOK FREEZER BLACKBERRY JAM

1 quart blackberries to yield  
  2 cups crushed berries  
4 cups sugar

1 package pectin  
2 tablespoons fresh lemon  
  juice

Wash and dry 2-cup or less plastic containers. Measure 2 cups crushed blackberries into large bowl. Measure sugar into separate bowl. Stir sugar into blackberries and set aside for 10 minutes. Empty contents of pectin package into small bowl and add lemon juice. Stir pectin mixture into blackberries. Stir constantly for 3 minutes. Fill all containers immediately to within one-half-inch of tops. Wipe off top edges of containers and quickly cover with lids. Let stand at room temperature for 24 hours. Store in freezer until opened. After opening, store in refrigerator up to 3 weeks.  
*Certo Recipe, General Foods Corp., Kankakee, Illinois*

## BLACKBERRY PIE

3 cups fresh blackberries  
1 cup sugar  
3 tablespoons cornstarch  
2 tablespoons lemon juice  
$1/2$ teaspoon cinnamon

$1/4$ teaspoon nutmeg  
dash of salt  
pastry for two-crust pie (see  
  juneberry recipes)  
3 or 4 pats of butter

Preheat oven to 450 degrees. Combine berries, sugar, cornstarch, nutmeg, cinnamon, lemon juice, and salt. Line pie pan with pastry, add filling, dot with butter, cover with top crust, and prick in several places. Bake 15 minutes; reduce temperature to 350 degrees and bake about 40 minutes longer (until the juice bubbles thick and crust is golden brown).

## OATMEAL BERRY MUFFINS*

*Muffins*

3/4 cup whole-wheat flour
3/4 cup all-purpose flour
1/2 cup rolled oats
1/2 cup brown sugar
2 teaspoons baking powder
1 teaspoon baking soda

1 cup fresh blackberries*
2 eggs
1/2 cup buttermilk
1/2 cup butter or margarine,
  melted

*Topping*

1/4 cup margarine, softened
1/4 cup rolled oats

1/4 cup brown sugar
1 teaspoon cinnamon

Preheat oven to 400 degrees. In large bowl mix flours, oats, brown sugar, baking power, and baking soda. Add berries. Stir to coat. In small bowl beat eggs with a fork, then beat in buttermilk and butter. Add to flour mixture, stir until just blended. Fill greased muffin tins two-thirds full.

Combine topping ingredients. Mix until crumbly. Sprinkle a small amount on each muffin. Bake 15 to 20 minutes or bake in a greased 9-inch square pan for 30 minutes.
*Janet Belanger, Buckfield, Maine*

*About any of the berries in the book would work for this muffin recipe, except chokecherry, wild grape, and beach plum.

## BLACKBERRY CORDIAL

8 quarts blackberries
2 quarts cold water
2 cups white sugar for every
  quart blackberry juice
1 tablespoon each, whole cloves
  and allspice

1/2 teaspoon grated nutmeg
1 cinnamon stick
2 cups (1 pint) brandy or
  whiskey for every quart
  syrup

1. In large enameled kettle simmer together blackberries and water until mushy.

2. For every quart of juice, stir in 2 cups sugar.
3. Tie spices into a little cotton bag and drop into the juice-sugar mixture.
4. Bring to a boil and boil 15 minutes.
5. When cool, add 2 cups brandy or whiskey to every quart syrup, pour into sterilized bottles, and cork securely. This will mellow. Store in cool, dark place.

Yield: 4 to 6 quarts
*Jean Ann Pollard,* The New Maine Cooking: Serving Up the Good Life

## HOBO COOKIES*

| | |
|---|---|
| 4 cups flour | 1 teaspoon vanilla |
| 1 cup shortening | 1/2 teaspoon salt |
| 1 cup butter (oleo can be used) | 3 egg yolks, lightly beaten |
| 1 cup milk | 1 12-ounce jar blackberry jelly |
| 1 cake yeast or 1 tablespoon dry yeast | or jam |

Preheat oven to 350 degrees. Combine flour, shortening, and butter and work together like pie dough. Set aside. Scald milk (cool to lukewarm), dissolve yeast in it, and let it rest for 10 to 15 minutes. Add vanilla, salt, and egg yolks.

Mix the yeast mixture into the flour mixture. Put sugar on a board as you would use flour to roll out pie dough. Roll 1/4-inch thick. Cut into 2-inch squares. Add 1 teaspoon or so of jelly or jam to the center of the square. Pinch the corners of the square together. Bake on greased baking sheets for 12 to 15 minutes or until dough just starts to brown on the edges.

Yield: 5 dozen cookies
*Emily Krumm, Eaton Rapids, Michigan*

*Any of the jams or jellies listed in the book are great in hobo cookies. Try a variety of flavors!

## BLACKBERRY JELLY

3 quarts blackberries
1/4 cup lemon juice

4 1/2 cups sugar
1 package pectin

Use fully ripe berries. Wash fruit. Crush, grind, or mash thoroughly. Heat to a boil. Rest colander in bowl or kettle. Spread cloth or jelly bag over colander. Pour hot prepared fruit into cloth or bag. Fold cloth to form bag and twist from top. Press with potato masher to extract juice.

Measure out 3 cups of juice. Put it, plus the lemon juice, in a 6- to 8-quart saucepan or kettle. Add pectin. Stir thoroughly to dissolve and scrape sides of pan to make sure all the pectin dissolves. Place over high heat. Bring to a boil, stirring constantly. Add sugar and mix well. Continue stirring and bring to a full rolling boil (a boil that cannot be stirred down). Boil hard exactly 2 minutes. Stir constantly to prevent scorching.

Remove jelly from heat. Skim foam and pour into sterilized containers. Fill jars to within one-eighth-inch of top, place two-piece metal lids on the jars and place in boiling water bath for 5 minutes.

*MCP Foods, Inc., Anaheim, California*

# Black Currants

Harvesting currants has largely been accidental for me. If I set out searching for another berry, like chokecherry or juneberry, I would stumble onto several loaded currant bushes. Conversely, when I went hunting specifically for currants, it seems they were nowhere to be found.

One year, when I wanted to make currant jelly and syrup, I headed for my favorite currant spot near a river. While there was an ample crop, I couldn't get enough for what I had in mind. You see, the currant bushes were heavily guarded by red ants. If I tried picking the currants, the ants attacked with such vigor that I had to back off. The ants prevented me from getting enough currants to put up a batch of jelly! Ants and other nasty insects are hazards that are seldom mentioned in field guides.

Some insect or other will be enamored of almost every kind of berry mentioned in this book. I've had close calls with bald-faced hornets, paper wasps, honeybees, yellow jackets, and ants. My advice is to keep your eyes open so you don't stumble onto a hive, a hill, or a nest of biting insects. No berry is worth the discomfort of an insect bite or sting.

The black currant is a common member of the genus *Ribes* — a genus that includes the currants and gooseberries. All members of the genus have an identifying characteristic: The dried remains of the flower stay attached to the top of the fruit—rather like a tail. (If you define the point where the stem attaches to the fruit as the bottom, the flower remains are at the opposite end, hence the discrepancy with the photo.) All members of the genus have leaves that are three-lobed like small maple leaves.

The black currant has a rather widespread distribution in New England. It blossoms in May. Its yellow, trumpet-shaped flowers have a pleasant fragrance of cloves.

Black currants ripen from late July through August. The size of the jet black fruits range from pea-sized to about twice the size of a pea. While there are other black berries that ripen in August, none has the dried, trumpet-shaped flower parts on the fruit that the currant has.

Currants grow as shrubs, that is, many-stemmed, fairly low-growing bushes that seldom exceed four feet in height. Black currant stems and branches lack thorns and are a glossy, deep brown to black color. They prefer deep, well-drained soils, but they are hardy and will grow in most areas except boggy sites. River floodplains are the ideal habitat for currants.

The black currant has a variety of uses, but its tart flavor makes it especially good for jelly and syrup. The taste certainly embellishes a piece of toast, a stack of pancakes, or waffles.

Currants are easily dried, and they can be used in recipes interchangeably with raisins. After washing and removing the dried flower parts and stems, I place them on a screen out-of-doors and let the sun do its work. If you prefer to get the job done faster, you can buy a commercial dehydrator designed specifically for drying fruits and vegetables.

Since black currants are an intermediate host for white pine blister rust, forestry departments have tried to eradicate this species and other members of the genus, but some of New England's more mountainous and remote areas still have good concentrations of currants. If you really wanted to pick currants, you should have been here 75 years ago. Today, currants are tough to find. You might have a one-in-three chance of finding enough currants for jelly (about a gallon). Keep your eyes open for currants while you're picking blueberries, juneberries, red raspberries, or wild strawberries. If you discover two or three bushes that are weighed down by a bumper crop of currants, pick them. If you only get a quart or so of black currants, don't despair, for currants can be mixed with other berries (gooseberries, red currants, yellow currants, chokecherries, and red raspberries) for dynamite jellies and syrups.

# Recipes

### OLD-FASHIONED BLACK CURRANT JELLY

1 gallon black currants
1 cup water

3/4 to 1 cup sugar per cup of currant juice

Wash currants and place layer of berries in 8-quart pot. Mash currants, add another layer, and repeat until pot is one-third full. Add water, cover, and heat to near boiling. Then simmer 30 minutes. Remove from heat, put mixture in wet jelly bag, and allow juice to drain for 2 to 3 hours. (Squeeze bag to hasten process, or force cooked berries through sieve.)

If you're daring, you can make jelly without pectin. Add 3/4 to 1 cup sugar per cup of juice (3 cups juice for a batch). Boil mixture 15 minutes, stirring constantly. Try jell test at this point. Dip clean metal spoon into boiling syrup, lifting spoon about 1 foot above kettle and allowing contents to drop back into syrup. When last few drops fall side by side or form a sheet, leaving spoon clean, jelly point has been reached.

Continue boiling jelly 1 or 2 minutes longer, then remove from heat, skim, and pour into sterilized jars. Seal with paraffin or securely tighten two-piece lids and place jars in boiling-water bath 5 minutes.

## BLACK CURRANT JELLY (USING PECTIN)

4 quarts currants to yield     1 package pectin
  5 cups juice                $7^1/_2$ cups sugar
$1^1/_2$ cups water

Wash and crush the currants in a kettle. Add water and simmer, covered, 15 minutes. Rest a colander in a bowl or kettle. Spread cheesecloth or a jelly bag to cover the colander. Pour in the hot currants. Fold cloth to form bag and twist from top. Press with potato masher to extract juice.

Measure 5 cups of prepared juice into 6- or 8-quart kettle or saucepan. Add pectin and mix thoroughly. Place mixture in kettle over high heat. Bring to a boil, stirring constantly. Add sugar and stir in. Continue stirring and bring to a full, rolling boil. Boil hard for 2 minutes. Remove from heat, skim off foam, and pour into sterilized jars. Wipe off rims. Use the USDA recommended boiling water bath method. Attach two-piece metal lids securely and place jars in boiling water for 5 minutes.

## CURRANT AND RASPBERRY JAM

2 cups crushed raspberries     2 cups crushed currants
  (about 3 pints)                3 cups sugar

Inspect, wash, thoroughly drain, and crush the raspberries. Wash the currants, remove stems, and crush. Put the raspberries, currants, and sugar into a saucepan over low heat and stir until the sugar is dissolved. (If scales are available, use 3/4 pound of sugar to each pound of fruit.) Increase the heat and boil rapidly, stirring constantly, until the juice passes the jelly test (the last drops of juice off the spoon form a sheet or drops fall side by side). Pour into hot jars and seal. Yield: three 8-ounce jars

## CURRANT WINE

8 quarts currants          2 gallons water
5 pounds sugar

Pour water over currants and sugar in enamel pan. Simmer over low heat until berries burst. Stir well and turn off heat. Cool to warm, then strain juice into crock. Fermentation should begin within 3 days. If it doesn't, stir in 3 packages of dry yeast.

*Phyllis Hobson,* Making Your Own Wine, Beer & Soft Drinks: A Garden Guide of Homestead Recipes

## JELLY ROLL*

1 cup flour                     4 eggs
1 teaspoon baking powder        1 teaspoon vanilla
dash of salt                    1 jar black currant jelly
1 cup sugar

Preheat oven to 375 degrees. Measure and sift dry ingredients into bowl. In another bowl, beat eggs and vanilla together, then add to the dry ingredients. Beat by hand until all the flour has been added. Pour onto a greased and floured cookie sheet. Bake for 20 minutes.

As soon as the roll is done, remove from the oven and peel back edges of roll from sheet. Invert roll onto a dish towel that has been sprinkled with powdered sugar.

Cut off crusty edges of roll. Stir up black currant jelly with a knife. Spread the jelly onto roll.

Roll up quickly, using towel as a guide. Remove towel and place jelly roll on a rack (seam side down) to cool.

*Emily Krumm, Eaton Rapids, Michigan*

*This recipe can be used with any jelly or jam; chokecherry, wild grape, blackberry, blueberry, and elderberry seem to be better choices than others. The black currant's strong, tart taste makes it ideally suited for jelly roll.

## CURRANT MEAD

2 gallons boiling water
9 pounds honey
2 pounds chopped, seedless
    raisins

1½ gallons currant juice
1 cup homemade liquid yeast
    (or 1 yeast cake dissolved in
    1 cup water)

To boiling water, add honey, raisins, and currant juice. Stir well to dissolve honey, then pour into a 5-gallon crock and cool to lukewarm. Add yeast and stir well. Cover and let stand in a warm (70 to 80 degrees) place 2 to 3 weeks.

When active fermenting stops, bottle, corking loosely, and stand bottles upright on a shelf in a cool storage room. Check every week to see if tiny bubbles have ceased, then cork tightly and lay bottles on their sides. Mead is ready to drink in 6 to 9 months.

*Phyllis Hobson,* Making Your Own Wine, Beer & Soft Drinks: A Garden Guide of Homestead Recipes

## BERRY APPLE CRISP

1 cup currants
2 medium apples, sliced
1 teaspoon cinnamon
¼ cup sugar

½ cup flour
⅓ cup powdered milk
¼ cup honey
¼ cup softened margarine

Preheat oven to 350 degrees. Combine currants and apples in a greased 9 x 9 baking dish. Sprinkle with cinnamon and pour sugar over all. In separate bowl combine flour, powdered milk, honey, and margarine. Mix until crumbly and sprinkle over fruit. Bake for 45 minutes.

*Janet Belanger, Buckfield, Maine*

## BERRY SYRUP

1 cup currants                    ¹/₄ cup water
¹/₃ cup honey

Combine currants, honey, and water. Crush fruit with back of spoon to release the juice. Boil 10 minutes or until it reaches a syrupy consistency.
*Janet Belanger, Buckfield, Maine*


## DANISH-STYLE RED CABBAGE AND CURRANTS

3 tablespoons margarine           ¹/₄ teaspoon cloves
1 red cabbage, shredded finely    1 cup currant syrup or more
1 teaspoon salt                       (use Berry Syrup above)

Melt margarine in heavy pan. Add finely shredded cabbage, salt, and cloves. Stir in currant syrup, cover, cook very slowly for 3 hours, stirring frequently. Add more syrup whenever the liquid simmers completely away—which will be several times. Serve *piping hot!*
*Janet Belanger, Buckfield, Maine*

# Beach Plums

Beach plums have an added attraction—they can offer the fun of a day spent along the ocean coupled with a berry-picking expedition. Those perfect September days along the Atlantic can only be enhanced by finding a clump of beach plums laden with fruit. It's great fun to take a picnic, laze on a sand dune, watch the gulls wheel overhead, soak up the sun, and then leisurely pick a peck or two of tart beach plums. Your beach plum expedition might be your last trip to the shore for the year, but it can be the most memorable.

Beach plums have an appropriate name for they are closely associated with beaches and dunes. Occasionally they are found inland growing in sandy areas, but most often you'll find this straggling shrub at the upper limits of beaches and sand dunes from southern Maine to Delaware.

Depending on the location, beach plums blossom from late April to early June. The plums turn a purplish color when they mature sometime between mid-September to mid-October.

While the shrub usually is low and straggling, sometimes it will be densely branched and reach a height of seven to eight feet. The leaves are oval, one to three inches in length and sharply toothed. The twigs are velvety—the buds, hairy. The undersides of the leaves are hairy.

Beach plums have a rather restricted habitat, but they can be plentiful in certain locations. If you find one of these spots, you can easily pick a gallon or two of the nickel- to quarter-sized plums. These slightly tart plums make terrific jam. You may consume the plums as is, but the tartness will probably make you stop after one or two. When they turn purple and soften, they are at their best for eating.

Like other wild plums, beach plums are very susceptible to late frosts. If the frosts hit while the plums are blossoming or just as they are setting fruit, the year's crop will be decimated. One year, when a severe freeze hit just as the plums blossomed, I searched more than two thousand plum bushes and found only one plum!

Plums will have peak years where every bush will be bent over. The next year, the same bushes might have only a handfull. Again, a little preseason scouting will help you locate the bountiful clumps of plums.

If you ranked all the berries and fruits in this book for the most pounds picked per hours spent picking, beach plums would rank number one. When you hit a good beach plum patch, you'll be able to pick a peck to a bushel in an hour or two. None of the other berries or fruits can come close to that time span and quantity.

Beach plums store easily. You can keep them in the refrigerator for two weeks or so, but it's easiest to freeze washed and drained plums in gallon-sized freezer bags. This won't hurt them for making jelly, jam, or cobbler. In fact, it's much easier to pit thawed plums than fresh ones. Commercial pitters, which are available from seed companies, specialty cookware places, and garden shops, make short work of the chore.

Let me warn you that beach plum jam and jelly are addictive. Make sure you have enough to last you from one September to the next. It's awful to run out of beach plum jam after you have become hooked on it!

# Recipes

## BEACH PLUM JAM

4 pounds beach plums to yield
  6 cups prepared pulp
1 cup water

1 package or pouch pectin
8 1/2 cups sugar

To obtain pulp, wash plums, then add 1 cup of water and simmer plums for 15 minutes or so, until cooked. Allow to cool, then squeeze each plum to remove pit. Save the juice and combine with the pulp.

Place the 6 cups of pulp in a 6- or 8-quart kettle or saucepan and mash with potato masher. Thoroughly stir in pectin and cook mixture over high heat. Stir constantly to prevent scorching. Bring to a boil. Pour in sugar and bring to full, rolling boil still stirring constantly. Boil exactly 4 minutes, remove from heat, skim off foam, and pour into sterilized jars. Wipe off rims and threads with damp cloth and securely fasten two-piece metal lids. Put in boiling water bath for 5 minutes.

Yield: ten 8-ounce jars

## BEACH PLUM COBBLER

3 cups beach plums, halved
  and pitted
3/4 cup sugar

1 tablespoon all-purpose flour
2 tablespoons butter
3/4 teaspoon cinnamon

*Biscuit dough*

1 3/4 cups sifted all-purpose flour
1/2 teaspoon salt
3 teaspoons double-acting
  baking powder

1 tablespoon sugar
4 to 6 tablespoons chilled
  butter
3/4 cup milk

Preheat oven to 425 degrees. Combine beach plums, sugar, and flour in saucepan and heat. Allow mixture to boil, then place it in 8 x 8 pan or casserole dish. Dot with butter and sprinkle with cinnamon. Prepare biscuit dough by sifting together flour, salt, baking powder, and sugar. Cut butter into dry ingredients, then add milk all at once. Stir until dough is fairly free from sides of bowl. Spoon dough over hot fruit. Bake about 30 minutes.

## BEACH PLUM JELLY

2 quarts beach plums (to yield approx. 4 cups juice)

$2^1/_2$ cups water
4 cups sugar

Cook beach plums in a pot with water and strain through a clean cloth to obtain clear juice. Mix the juice and sugar well and bring to a boil. Cook, stirring constantly, until mixture reaches the syrup stage (when a drop of the hot juice dropped into a saucer of cold water has a jellylike consistency). Cook for a few more minutes, remove from heat, and skim off any foam. Pour the jelly into scalded jelly glasses or jars. Cover with melted paraffin wax.

This is the old-fashioned way of cooking beach plum jelly. It tends to become too solid if overcooked.

**NOTE:** *Never* heat paraffin over a direct flame—always put paraffin in a double-boiler or in a saucepan over a pan of hot water.

*Mary Alice Cook,* Traditional Portuguese Recipes from Provincetown

## BEACH PLUM JELLY USING PECTIN

2 quarts beach plums yielding     6 cups sugar
   3¹/₂ to 4 cups juice           ¹/₂ bottle Certo
2¹/₂ cups water

Place 2 quarts of beach plums in a pot with water. Cook for about 25 minutes, mashing the plums carefully so as not to get burned by hot, splashing juice. Strain mixture through a clean cloth. Do not squeeze the cloth or the jelly will become cloudy. Just leave it to drip slowly.

Mix juice and sugar and bring to a boil. When a rolling boil is reached, add the Certo. Boil hard for 2 minutes, stirring well. Remove the jelly from the heat and skim off any foam. Pour hot jelly into scalded glasses or jars and seal with paraffin wax.

Leave the jelly to set until it is cold. Wait a day or two before using.

*Mary Alice Cook,* Traditional Portuguese Recipes from Provincetown

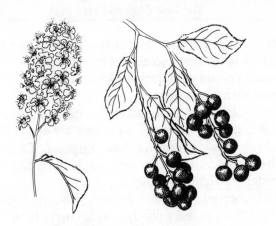

# Chokecherries

Chokecherries are widely distributed in New England, but not especially popular. Too bad, for a number of tasty treats can be concocted from the juice of these "pucker" berries.

For the past fifteen years, chokecherries have been part of my life, but I never picked them when I was a kid. I guess my mother figured that there were plenty of pie cherry trees around so why mess with such a puckery cherry?

Well, I found out that chokecherry jelly was not puckery, but full of tart taste and is so good that I started picking the cherries. To my astonishment, I found that I was not alone. Quite a few folks like chokecherries. Not only do people like chokecherries, but the wildlife does, too. Chokecherries are relished by birds, raccoons, bear, deer, and wild turkey.

My sons, James and Clint, love chokecherry jelly, but like all teenagers, don't particularly like to work—especially without pay. The one summer that I didn't have time to pick chokecherries was followed by a winter without chokecherry jelly. The boys started protesting that they hadn't any jelly for their toast, nor any tasty syrup for pancakes and waffles. I reminded them that they had no chokecherry jelly because they hadn't picked any chokecherries. You can bet they picked a gracious plenty the following August. To top it off,

James and Clint even rendered the cherries into juice and made four of the seven batches of jelly. In the years since, the boys have continued to pick chokecherries on their own and to make their own jelly.

The chokecherry blossoms from late May to early June. The long, graceful, pale-white racemes are one of the prettiest shrub flowers gracing New England. Chokecherries mature in August, when the dark purple-black cherries occur in clusters of eight to twenty.

Chokecherries prefer well-drained, rich soils, but will grow in all but the soggiest soils. The chokecherry is an invader—that is, it grows in previously disturbed areas. The cause of the disturbance could be fire, farming, road or railroad construction, or any other force that removes the trees from an area.

Chokecherries need a good dose of sunlight. Heavy shade will eventually eliminate chokecherry from a site.

Chokecherries can be thought of as large shrubs for they grow as high as twenty feet, but most attain a height of eight to ten feet.

The bark of chokecherry has a dark gray hue with noticeable lenticels, that is, breathing pores. The leaves are elongated oval, with a pointed tip and serrated (toothed) margin. The leaves are arranged in alternate fashion.

The chokecherry fruit is called a drupe and has a thin skin over a hard pit. Only the skin and its juices are used in any concoction, so large amounts of chokecherries are needed for making jelly or syrups. (Plan on picking at least two gallons of chokecherries for a batch of jelly or syrup.)

A note of caution about chokecherries: The kernel in the pit contains cyanogenetic glycosides, which turn into hydrocyanic acid. The leaves, bark, and stems of this shrub also have hydrocyanic acid. When you cook the chokecherry to render the juice, the cyanide compound is destroyed. Only the flesh of the chokecherry is edible. Do *not* eat any other parts of this shrub!

Chokecherries are easy to pick. I've picked five gallons in a little over an hour. It's best to pull the cherries off the stems while picking. I like to pick into a berry bucket made from a gallon milk jug with part of the top cut out. I leave the handle on the jug so I have a convenient hold. I can wade into a clump of chokecherries with my milk jug, pull down a laden branch, strip off the cherries, pull down another branch, pick it clean. I keep it up until I've either picked all the chokecherries or I have enough.

By the way, if you pick too many to render within a day or two, wash them, let them drain dry, put them in gallon plastic bags, and place them in the freezer. They'll keep for six months or so and actually be easier to render than fresh chokecherries.

Tent caterpillars like chokecherry leaves. The silky tents the caterpillars build are very visible, so if you're having trouble finding chokecherries, you might look for tent caterpillars instead. Once you locate the chokecherries via tent caterpillars, you will no doubt find some bushes that are not infested by the caterpillars.

Since birds, mammals, and tent caterpillars like chokecherries, sometimes they take the lion's share of the fruit. The tent caterpillars will so weaken the plant that it won't produce much of a crop or the silk will make it impossible for you to pick the cherries. Most years, though, you will have an excellent chance of getting enough chokecherries for at least a batch or two of jelly.

# Recipes

Chokecherries are too tart and have too large a pit to eat as is. The best uses of chokecherry come from its juice. In other words, chokecherry makes great jelly, syrup, and wine.

To get chokecherry juice, place a layer of the washed cherries in the bottom of an 8-quart or larger pot. Mash. Add another layer of cherries and mash. After you've repeated this process until the pot is one-third to half full, add 2 cups of water, cover, and heat to near boiling, then simmer for a half hour or so. Stir occasionally.

Place cooked chokecherries in a large bowl to cool. Render the juice by forcing the cooked berries through a colander or sieve, or by placing them in a wet jelly bag; suspend bag, and let juice drip into receptacle placed underneath. You can also squeeze the jelly bag to get the juice out quicker.

Another method is to use a strainer (Foley Food Mill) and to add a half cup of cooked chokecherries at a time to the strainer. Run over the chokecherries a dozen or so times, then discard the pits and make another batch.

## CHOKECHERRY WINE

| | |
|---|---|
| 1 gallon chokecherries | 3 pounds sugar |
| 1 gallon water | 1/2 cake fresh yeast |

Combine fruit, water, 2 pounds of sugar, and yeast in stone crock or glass container. Let stand in dark place at room temperature for 2 weeks, stirring once a day. Keep covered with dish towel. After 2 weeks, remove chokecherries and add remaining 1 pound of sugar and set aside for another 2 weeks, stirring once a day. Keep covered. After 2 weeks, siphon off juice and bottle.

*Dwight Layton, Worcester, Massachusetts*

## CHOKECHERRY SYRUP

4 cups chokecherry juice      2 cups white corn syrup
4 cups sugar

Place all ingredients in 5-quart kettle. Bring to a boil, stirring to dissolve sugar. Turn heat to moderate and continue boiling for 10 to 15 minutes or until foam starts to climb sides of kettle. Remove from heat and pour into sterilized jars, and seal. After opening, syrup will keep in refrigerator for weeks.
*Marjorie Helms and Alice Halsted, Sheridan, Wyoming*


## CHOKECHERRY JELLY

3 cups chokecherry juice      1 bottle Certo pectin
6½ cups sugar

Measure 3 cups of chokecherry juice into 6- or 8-quart kettle or saucepan. Add the sugar and mix well. Place over high heat and bring to a boil, stirring constantly. At once, stir in Certo. Then bring to a full rolling boil and boil hard 1 minute, stirring constantly. Remove from heat, skim off foam with metal spoon, and pour quickly into sterilized glasses. Cover at once with one-eighth-inch hot paraffin or use USDA recommended boiling water bath method.
Yield: 6 cups jelly
*Susan Littlefield, Topsham, Maine*


## CHOKECHERRY LIQUEUR

Layer chokecherries and 1 to 1½ cups sugar in quart jar. Cover jar with loose-fitting lid. Fill to top with brandy and let set for 2 to 3 months, then siphon off.
*Dwight Layton, Worcester, Massachusetts*

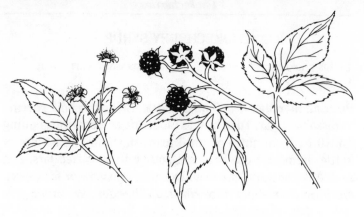

# Black Raspberries

Some of my best berry-picking memories come from picking black raspberries. I grew up in a farming area checkered with woodlots and cut by numerous creeks and rivers. Many of the river and creek floodplains had deep, loamy soils. It was an easy matter to walk along the rivers and pick a two-gallon bucketful of black raspberries in an hour or two. The berries grew in such abundance that even after I ate a pint or more, I had plenty left for the berry bucket.

Although there were plenty of black raspberries growing in the abandoned fields, the sun dried out the berries, so I avoided those areas. Besides, I was much cooler walking in the open woods along the Grand River. Black raspberry-picking expeditions were also reconnoitering trips. While I was picking black raspberries, I could also gauge the quality of the blackberry crop and figure out how the pheasants, cottontail rabbits, and ducks were doing.

Since black raspberries grew in such profusion, they were the only berries my mother ever asked me to quit bringing home. After she had frozen twenty quarts, put up four batches of jam and jelly, and made six pies, she would beg me to stop picking black raspberries. Perhaps it was for the better because I picked so many of the perfectly ripe berries that my hands were stained a deep purple. It took a couple of weeks to lose the stains.

Black raspberries are great for trail snacks. They also can't be beat with just a sprinkle of sugar and cold milk or as a topping for cold cereal or ice cream. My mouth starts watering whenever I think of my mother's black raspberry pie. Black raspberries make some of the best jam and jelly going.

One of the tastiest berries to grace New England, the black raspberry looks a lot like the red raspberry but with a few differences. Of course, the ripe fruit is a deep purple-black color, sometimes with a light bloom. The canes (stems) look like red raspberry canes, except that they are curved. Often the cane tips will touch the ground and root. The canes have a purplish hue with a heavy white bloom. In general, black raspberry canes have fewer thorns than red raspberry, though the black raspberry thorns tend to be larger than red raspberry ones.

Black raspberries are spread by birds. They'll ingest the berry but can't digest the seeds. Wherever the birds defecate the seeds, black raspberries are likely to germinate and prosper.

With pale, white blossoms similar to red raspberries, black raspberries blossom in mid- and late May and mature in mid-July and last into August. Initially the berry turns a bright glossy red, then purple, and finally a lustrous purple-black. As with red raspberries, the receptacles remain on the bush when you pick them. Black raspberries have attained the peak of flavor when you can just touch the berry and it falls into your hand.

Black raspberries need sunlight, hence you will find them growing in abandoned fields and along fence rows and rights-of-way. The best places to find tasty black raspberries are on deep, well-drained soils, particularly stream floodplains. Floodplains usually have deep, loamy soils with some tree cover that offers moderate shade and allows the berries to ripen slowly. Consequently, they are succulent, attain a big size, and have plenty of taste! Those black raspberries growing in open sunlight often dry out and end up more seed than flesh.

The leaves are compound with normally three or sometimes five light-green—hued leaflets. The two or four lateral leaflets lack stems while the terminal leaflet has one. The undersides of the leaflets have hairs. The leaflets are palmately arranged, i.e., all come from one point on the leaf stalk.

In southern New England, you stand an excellent chance of finding ample amounts of black raspberries, but the northern reaches don't have such widespread concentrations of the fruit. The scattered pockets there might limit your using them only for trailside snacks or garnishes.

Black raspberries are easy to keep. Just rinse them in cold water, drain thoroughly, place them in plastic freezer bags, and freeze. They will keep their freshness for about six months.

While you're picking black raspberries, look for other ripe berries. Juneberries and currants will be ripening at about the same time. Also, you should be able to do some scouting for blackberries and elderberries. Just make sure to enjoy some of the black raspberries as you wander about.

Blueberry twigs in winter.
Note terminal bud and that the
buds are alternate.

Blueberry blossoms.

Huckleberry blossoms.

PLATE 1

Blue blueberries. Note that crown (calyx) is five-pointed.

Black blueberries. Note that crown (calyx) is five-pointed.

Blackberry blossoms.

Dewberry blossoms.

PLATE 2

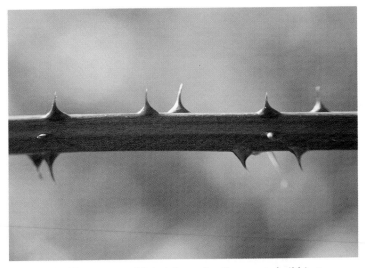

Blackberry cane. Note triangular thorns and ribbing.

Ripe blackberries.

Ripe, ripening, and green dewberries.

PLATE 3

Currant blossoms.
Note trumpet-shaped flower.

Ripe currants. Note that
flower parts are still attached.

Ripe currants. Note deeply lobed, maplelike leaves.

PLATE 4

Beach plum blossoming on sand dune.

Closeup of beach plum blossoms.

Ripe beach plums. Note powdery bloom on plums and hairy undersides of leaves.

PLATE 5

Chokecherry twig. Note
prominent lenticels.

Chokecherry blossoms. This type of flower is a raceme.

Green chokecherries.

PLATE 6

Ripening chokecherries.

Ripe chokecherries.

PLATE 7

Black raspberry canes. Note purplish bloom and downward curve of cane.

Ripe and ripening black raspberries. Note receptacle— it remains on the bush when you pick a raspberry.

Another view of ripe and ripening black raspberries.

PLATE 8

Cranberry blossoms.

Ripe large cranberry. Note small leaves and short bush.

Mountain cranberry blossoms.

Ripe mountain cranberry.

PLATE 9

Elderberry blossoms.

PLATE 10

Ripe elderberries.

Another view of elderberry bush with ripe
and ripening elderberries.

Elderberry twig. Note prominent
lenticels.

Cross section of elderberry
stem. Note white pith.

PLATE 11

Typical grape leaf with damsel fly on it.
Note leaf's large lobes and teeth.

Grapevine showing autumn color.

Ripe grapes. Note that grapes
are in clusters and have a powdery bloom.

PLATE 12

Wild grape climbs very well.

PLATE 13

Juneberry blossoms. This species blossoms
before the forest greens.

Another species of juneberry.

Ripe and ripening juneberries. Note
that crown (calyx) is five-pointed.

Ripening juneberries.

PLATE 14

Ripe raspberries. Note compound leaf with large teeth.

Raspberry cane with fine prickles.

Another raspberry cane with sparse but good-sized thorns.

PLATE 15

Green strawberries. Note lack of woody stems.

Ripe strawberries. Note compound leaf with large teeth.

PLATE 16

# Recipes

## BLACK RASPBERRY PIE

4 cups black raspberries
1 tablespoon lemon juice
4 tablespoons flour
pastry for 2-crust pie (see
  juneberry recipes)

1 cup sugar
1/2 teaspoon cinnamon
  (optional)
2 teaspoons tapioca

Preheat oven to 450 degrees. Roll out the bottom pie crust and place in pie pan. Mix the pie ingredients together and place in pie pan. Add 2 to 4 pats of butter. Cover with top pie crust. Place on a cookie sheet and put in oven. Bake for 10 minutes, then lower heat to 350 degrees and bake until done (30 minutes or so).

*Emily Krumm, Eaton Rapids, Michigan*

## BLACK RASPBERRY COBBLER

2 cups black raspberries
1/2 cup sugar

1 tablespoon tapioca

*Cobbler*

1 1/2 cups sifted flour
1/2 teaspoon salt
2 teaspoons baking powder
1/2 cup sugar

1/4 cup butter
1/3 cup milk
1 beaten egg

Preheat oven to 450 degrees. Place berries, sugar, and tapioca in an 8 x 8 baking dish. Sift dry ingredients together. Cut in butter. Combine milk and egg. Add to dry ingredients. Stir together just until all flour is moistened. Drop batter in six mounds onto black raspberry mixture. Place in oven and bake for 15 minutes, then lower heat to 350 and bake for 30 minutes more. Serve with vanilla ice cream.

*Emily Krumm, Eaton Rapids, Michigan*

## BLACK RASPBERRY JELLY

3 quarts black raspberries to
   yield 4 cups prepared juice
1 cup water

5$^1$/$_2$ cups sugar
1 package pectin
$^1$/$_4$ cup lemon juice

Wash and mash berries thoroughly. Add water and lemon juice. Bring to a full rolling boil. Remove from heat. Rest colander in bowl or kettle. Spread cloth or jelly bag over colander. Pour hot black raspberries into cloth or bag. Fold cloth to form bag and twist from top. Press with potato masher to extract juice.

Measure sugar into dry bowl and set aside. Measure 4 cups of juice and put in a 6- or 8-quart kettle or saucepan. Add package of pectin and stir thoroughly until dissolved.

Place mixture in kettle over high heat. Bring to a boil while stirring constantly. Add sugar and mix well. Continue stirring and bring to full rolling boil (one that cannot be stirred down). Boil hard for exactly 2 minutes while stirring constantly. Remove jelly from heat. Skim foam and pour into containers. Fill jars to within one-eighth-inch from tops, and clean jar rims and threads. Put on two-piece metal lids. Secure bands firmly. Place in boiling water bath for 5 minutes.

Yield: six 8-ounce jars

*MCP Foods, Inc. Anaheim, California*

## BLACK RASPBERRY JAM

8$^1$/$_2$ cups sugar
3 quarts black raspberries

1 package pectin
$^1$/$_4$ cup lemon juice

Measure sugar into dry bowl and set aside.

Wash and grind berries thoroughly until reduced to pulp. Measure 5$^3$/$_4$ cups pulp and put it and lemon juice in a 6- or 8-quart kettle or saucepan. Add pectin and stir thoroughly until dissolved.

Place mixture in pan over high heat. Bring to boil, stirring constantly to prevent scorching.

Add sugar and mix well. Continue stirring and bring to full rolling boil (a boil that cannot be stirred down).

Boil hard exactly 4 minutes. Stir constantly to prevent scorching.

Remove jam or jelly from heat. Skim foam and pour into containers. Clean off jar rim and threads and put on two-piece metal lids. Screw bands on tightly. Process in boiling water bath for 5 minutes.

Yield: ten 8-ounce jars

*MCP Foods, Inc., Anaheim, California*

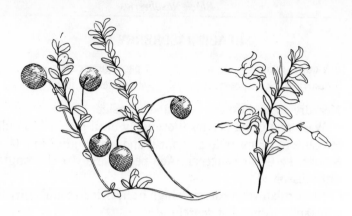

# Cranberries

I grew up in Michigan where cranberries are not nearly as common as in New England. In fact, I never saw a wild cranberry before 1987. For some reason, I assumed that cranberries were only grown domestically and did not occur in the wild. Imagine my surprise when my friend Jan White of Woolwich, Maine, told me there were oodles of cranberries growing wild in Maine and throughout most of New England.

Well, I bought a field guide and attempted to find them. The book's photos and descriptions misled Dot and me. We were looking for a bush about two to three feet tall with half-inch long leaves. We searched several bogs and found no cranberries. We drove to Jan's to ask for his help.

He said, "Go out of my driveway, go left for 250 yards, park by the broken oak tree, cross the road, and walk down the stone fence that goes into the marsh. When you reach the marsh, you'll be in a mess of cranberries."

I followed Jan's directions and was ankle deep in sphagnum moss, but I didn't see any bushes two to three feet high. About that time, Dot said, "Here they are."

I told her that I didn't see anything. She exclaimed, "Look down at your feet."

I did and, sure enough, there they were. What surprised

me was that the cranberry bushes were just straggly stems bearing minuscule leaves. The bright, red cranberries were startlingly large by comparison. Each little twig had one or two nearly marble-sized berries on them.

Dot and I discovered that the ripe cranberries were puckery, but still a lot sweeter than the commercial varieties. We also discovered that we weren't properly outfitted in order to pick cranberries. We didn't have a decent berry bucket, and we certainly didn't have proper footgear. Field guides never mention that it's downright wet where lowland varieties of cranberries grow. If you go to a cranberry bog on a blustery November day without the proper footgear, expect to get cold, wet feet! To pick cranberries in comfort, wear knee-high rubber boots and a warm jacket.

By the by, the cranberries we picked did make delicious Thanksgiving treats—cranberry bread and the White family recipe for cranberry sauce. Try the recipes, you'll love them.

The later in the season you can pick cranberries, the riper they will be. Cranberries harvested in November have lots of natural sugars. They require much less sugar to make relishes and conserves than is needed for the commercial varieties that are picked in early October. While cranberries are grown commercially and are vital to the New England economy, the wild varieties are just as big and tastier than those you find on the grocers' shelves.

Cranberries belong to the same genus as blueberries, *Vaccinium*. Cranberries, as with other members of the genus, prefer acid conditions like sphagnum bogs.

Three species of cranberries occur in New England: mountain, large, and small cranberry. Large and small cranberry prefer bogs, but the large cranberry occurs in a greater variety of places. It grows in bogs, swamps, on wet shores, and in low meadows and even on humus-covered rocks along the coast. The small cranberry prefers sphagnum and wet humus. Mountain cranberry grows on cool, moist woodlands on rocky mountainsides.

All three species blossom in May and mature in October. Large cranberry often holds its fruits through Thanksgiving and mountain cranberry even holds its fruits into spring.

One notable clue for identifying cranberry in the field is the small leaf size—3/8-inch long and barely 1/4-inch wide. The branches are thin and spindly. They reach less than a foot in height, and are prostrate. The leaves are deep green on top and pale, silver-green underneath. After a heavy frost, the leaves may develop a tinge of red.

Cranberries need fresh water to thrive. A salt marsh won't suffice for them, but you might find cranberries growing along the edge of a salt marsh where a stream or spring enters.

The fruits of large cranberry are round and resemble small Christmas ornaments. They are roughly the same size as the commercial varieties. The ripe cranberry has a color ranging from bright to deep red.

Cranberries are hardy, but the fruits are frost sensitive until they are ripe. An early frost will "blanch" the green berries. My friend Sam Ristich, a retired botanist, has a cranberry patch in a depression of a large blueberry field. He is able to harvest only one crop every four years because of the early frosts. He says that cranberries growing on wetter sites are protected by the water and aren't as adversely affected by the frosts.

Cranberries can be afflicted with a blight or a disease that prevents the berries from forming. The affliction cuts down the crop considerably.

Sam Ristich contends that the hardest part about cranberry picking is finding where they are. Once you locate a good cranberry bog, it's a cinch to pick two to four quarts.

If you get a good harvest of cranberries and can't use them all immediately, rest assured that they keep for quite a long time in the refrigerator. They also can be frozen—just wash and drain the berries, put them in plastic bags, and toss them into the freezer. They should keep well for six months.

Since this berry is available through Thanksgiving, it's fitting that there is a cornucopia of recipes for this tart, wild fruit.

# Recipes

### CRANBERRY SURPRISE

| | |
|---|---|
| 1/3 cup butter (or margarine) | 2 1/2 teaspoons baking powder |
| 1/2 cup sugar | 1/3 cup milk |
| 2 eggs | 1 cup cranberries (picked in |
| 2 1/3 cups flour | early November) |

Blend softened butter with sugar until creamy. Beat eggs. Add to sugar mixture. Mix dry ingredients together, sift. Add one-third dry ingredients to sugar along with one-third milk. Blend. Mix in remaining thirds. Stir in cranberries. Place in a double boiler and steam for 3 hours. Serve with cream seasoned with nutmeg.
*Dot Heggie, Falmouth, Maine*

### WILD CRANBERRY SHERBET

| | |
|---|---|
| 4 cups cranberries | 1/8 teaspoon salt |
| 1 cup boiling water | 2 egg whites |
| 2 cups sugar | 1/2 cup orange juice |

Cook cranberries in boiling water 10 minutes. Add sugar, salt, and orange juice. Run through sieve, cool. Turn into refrigerator tray and partially freeze. Remove from refrigerator, add egg whites, beat 1 minute. Return to freezer and freeze until firm.
*Grace Hoar, Grand Lake Stream, Maine*

## CRANBERRY BREAD

3 cups flour
4 teaspoons baking powder
2 cups sugar
1 teaspoon salt
1 cup milk

1 cup halved cranberries
1 tablespoon orange or lemon
rind
1/2 cup walnuts (optional)

Preheat oven to 325 degrees. Mix together first four ingredients. In separate bowl, combine egg and milk and add to dry ingredients, stirring well. Then add cranberries, orange or lemon rind, and walnuts (if desired). Place in two greased and floured loaf pans and bake 1 hour.
Yield: two medium-sized loaves
Cooking to Beat the Band. *Compiled by Band Mothers Club, Deering High School, Portland, Maine*

## CRANBERRY ORANGE BREAD

1 cup all-purpose flour
1 cup whole-wheat flour
1/2 teaspoon salt (optional)
1/2 teaspoon baking powder
1/2 teaspoon baking soda
1/2 to 1 cup sugar (to taste)
1 orange

3/4 cup orange juice
3 tablespoons vegetable oil
1 egg
1 cup chopped nuts
1 cup chopped raw cranberries
1 cup chopped nuts

Preheat oven to 350 degrees. Sift flours. Measure and sift with salt, baking powder, baking soda, and sugar. Set aside. Grate rind of orange (to produce 4 teaspoons). Squeeze out juice. Add enough additional orange juice to make 3/4 cup. Add vegetable oil to cup with orange rind.

Beat egg. Add orange mixture to it. Stir in sifted dry ingredients. Fold in chopped nuts and cranberries.

Pour batter into well-greased loaf pan. Bake 1 hour.

## CRANBERRY BETTY

2 cups cranberries
1 cup water
1 cup sugar

2 cups soft bread crumbs
2 tablespoons butter
1/2 cup seedless raisins

Preheat oven to 375 degrees. Cook cranberries, water, and sugar 10 minutes. Set aside. In a buttered baking dish, place layer of bread crumbs, then a layer of raisins, and one-half of the stewed cranberries; dot with 1 tablespoon butter. Repeat and cover top with bread crumbs, dotting again with butter. Bake until brown (about 30 minutes). Serve with whipped cream.

The Betty may be covered with meringue made from 2 egg whites, 4 tablespoons sugar, and 1/2 teaspoon vanilla; return Betty to oven for about 15 minutes or until meringue is delicately browned and cooked through.
*Dot Heggie, Falmouth, Maine*

## CRANBERRY SALAD

1 quart cranberries
2 1/2 cups boiling water
2 cups sugar
pinch of salt
2 tablespoons unflavored gelatin
1/2 cup cold water

1 package lemon-flavored Jell-O
1 cup boiling water
2 cups crushed pineapple
2 cups Tokay grapes, skinned
  and seeded

Cook cranberries in water until they burst. Press through sieve. Add sugar and salt. Soften unflavored gelatin in cold water. Add to hot cranberry juice. Dissolve Jell-O in boiling water. Add to cranberry juice. Stir thoroughly. When cool, add crushed pineapple, Tokay grapes. Chill several hours before serving.
Cooking to Beat the Band. *Compiled by Band Mothers Club, Deering High School, Portland, Maine*

## CRANBERRY WARBLER

1½ cups ground cranberries
½ cup sugar
1 package orange-flavored Jell-O
1 package lemon-flavored Jell-O
1 tablespoon lemon juice

⅛ teaspoon cloves
¼ teaspoon cinnamon
1 orange, diced
1½ cup chopped walnuts

Mix together cranberries and sugar and set aside. Combine two Jell-Os and prepare according to package instructions. To hot Jell-O, add lemon juice, ground cloves, and cinnamon. Just as Jell-O begins to cool, add cranberry mixture, orange, and chopped walnuts.

*Margaret Bourassa, Dixfield, Maine*

## PARADISE JELLY

4 quarts red apple or crab apple
12 quinces
water to cover

2 quarts wild bog cranberries
1 quart water
1 pound granulated sugar

Wash apples and quinces. Remove stems and blossom ends. Cut to same size pieces, cover with cold water. Cook until tender.

Wash cranberries and cook in 1 quart water until tender. Put apple, quinces, and cranberries into a jelly bag to let drain *overnight*. The next day, in a saucepan, add 1 pound white sugar for every 2 cups of juice. Boil about 10 minutes, stirring constantly. Remove scum. Pour into glass jars and seal with paraffin. (USDA recommends boiling water bath method.)

*Susan E. Littlefield, Topsham, Maine*

## CRANBERRY CONSERVE

1 orange
1 cup water
1/2 cup sugar
1 pound cranberries

2 cups water
2 cups sugar
1 cup seeded raisins
1/2 cup walnuts

Cut up the orange and its rind in small pieces and remove the seeds. Add 1 cup of water and 1/2 cup sugar and simmer until the rind is tender. Cook cranberries in 2 cups of water. Add 2 cups sugar, then the raisins. Continue to cook until the mixture thickens, then add the orange rind and the nuts. Pour into jars and seal.
*Adapted from Clarissa M. Silitch,* The Forgotten Arts, Making Old-Fashioned Jellies, Jams, Preserves, Conserves, Marmalades, Butters, Honeys & Leathers

## CRANBERRY PIE

3 cups cranberries
2 tablespoons flour
1 cup raisins
3 tablespoons water

1 1/2 cups sugar
1/4 teaspoon salt
1 tablespoon melted butter
pastry for two-crust pie (see
    juneberry recipes)

Preheat oven to 450 degrees. Wash cranberries. Chop and mix with flour, raisins, water, sugar, salt, and butter. Line pie pan with pastry, pour in filling, and cover with pastry. Bake 15 minutes; reduce heat to 350 degrees and bake about 30 minutes longer.
Yield: one 9-inch pie
*Margaret Bourassa, Dixfield, Maine*

## CRANBERRY APPLE PIE

2 tablespoons minute tapioca
1 1/2 cups sugar
1/4 teaspoon salt
Pastry for 2-crust pie (see
juneberry recipe)

2 1/2 cups diced, peeled, fresh
apples
1 1/2 cups ground cranberries
1 tablespoon butter or
margarine

Preheat oven to 425 degrees. Combine tapioca, sugar, salt, and fruits. Let stand 15 minutes. Line pie plate with bottom crust. Add fruit. Dot with butter and cover with top crust. *Seal edges.* Bake 15 minutes; then lower temperature to 350 degrees and bake for 40 minutes or until browned.
*Janet Belanger, Buckfield, Maine*

## MOCK CHERRY PIE FILLING

1 quart cranberries
1 cup raisins
1/2 cup water
3 tablespoons cornstarch
1/2 teaspoon salt

2 cups sugar
2 teaspoons vanilla
pastry for two-crust pie (see
juneberry recipes)

Preheat oven to 425 degrees. Cook cranberries and raisins in water 5 minutes. Then add cornstarch, salt, sugar, and bring to a boil. Add vanilla and cool. Prepare pastry for two pies. Line pie plates with pastry. Fill with cooked fruit, cover with pastry, and bake at 425 degrees 10 minutes; continue baking at 350 degrees until done.
Yield: 2 pies
*Cynthia Moses, Grand Lake Stream, Maine*

## MOLDED CRANBERRY SALAD

1 package cherry Jell-O
1 cup hot water
1/2 cup pineapple juice
1/2 cup sugar

3/4 cup chopped celery
1 cup chopped cranberries
1 cup chopped whole orange
1 cup mayonnaise

Add water to Jell-O, let stand until cool. Add pineapple juice and sugar. When thick, add remaining ingredients. Put in individual molds and chill.

Cooking to Beat the Band. *Compiled by Band Mothers Club, Deering High School, Portland, Maine*

## CRANBERRY SAUCE

4 cups cranberries
1 1/2 cups water

2 cups sugar
1/2 an orange rind, grated

Combine ingredients in saucepan. Bring to boil, then simmer until berries pop. Optional ingredients—cinnamon and nutmeg, a diced apple such as a spy, baldwin, winesap, or streak (an old Maine apple), walnuts or pecans. The optional ingredients can be added after the berries pop.

*White Family Recipe, Georgetown, Maine*

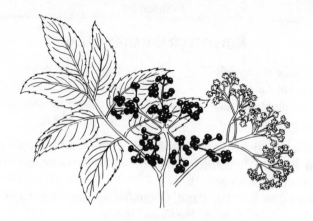

# Elderberries

When I was growing up, we went after elderberries just before the hard autumn frosts. It was a pleasant time to go berry picking for the mild days and cool evenings were just right for being out-of-doors. The muck lands near our Michigan home were ideal for elderberries.

My mother would take along two or three grocery bags. My sister, brother, and I would pick some of the lower branches while Dad and Mom would pick the higher branches. We would pick the berry clumps, go home, and sit down to pull off each berry. It was rather like shelling peas.

While we were out and about, we'd gauge the hickory nut and black walnut crops. We'd make mental notes of any heavily laden trees, so we could return after a heavy frost caused the nuts to fall.

Mom would waste little time with the elderberries. In a day's time she would have made a pie or two and put up the rest as jelly. The jelly would most certainly grace our table at Thanksgiving (as well as black raspberry, red raspberry, and blackberry jam or jelly).

When I think of elderberries, I think of delicious jelly, warm early autumn days, and our family putting up nature's bounty for the long winter ahead. I think of digging potatoes and carrots and storing them in the root cellar. I think of sauerkraut fermenting in the porcelain crocks also in the root cellar. I think of a larder full of canned corn, succotash, waxed beans, green beans, and pickled beets. I recall bringing in bushels of tomatoes to be canned as stewing tomatoes, chili sauce, and tomato juice. Elderberry time was hectic, but so full of bounty and blessings.

Elderberry blossoms in June. The small, white flowers are arrayed in flat, saucer-shaped clusters that look like an open umbrella—hence the term umbeliform, which refers to this type of flower arrangement. (Botanists call the arrangement a cyme.)

While most of the berries described in this book mature during the summer, elderberry ripens in the fall—late September through early October, to be exact. The umbrella-shaped clump of BB-sized fruits reaches a diameter of about six inches.

Elderberry bushes prefer deep, well-drained, rich soils. They prefer open to subdued sunlight and don't do well in deep shade. It's not unusual to find them growing along streams. While elderberries are quite common in most of New England, they are sparse in northwestern Maine.

Elderberries have opposite, pinnately compound leaves (the leaflets originate at different points) with three to five pairs of toothed leaflets with an odd one at the tip. The bush attains a height anywhere from three to thirteen feet. The canes (stems) have a woody exterior with prominent lenticels and a pithy or hollow interior. The bush usually covers a circular area with numerous canes in the clump. An average clump covers ten to forty square feet.

The easiest way to harvest elderberry is to pick the entire fruit cluster. You can pull the individual berries off when you reach home and can do the chore sitting down.

Elderberry has a mild flavor that makes for tasty jelly, delicious pies, and wines. The leaves, stems, and roots contain cyanogenetic glycosides so only the cooked berries should be eaten. Consuming a large number of the uncooked berries can have a laxative effect.

Birds relish elderberries. Seldom will you be able to pick a clump of elderberries that hasn't been picked over by birds. They won't eat all the berries, but at times, there won't be enough left in a clump for you to make a pie. I'd rate your chances of finding enough (2 cups) for a pie at 90 percent; for jelly, your chances are at 75 percent; for wine (20 pounds), 25 percent.

# Recipes

### ELDERBERRY JELLY

3 pounds elderberries
1 package dry pectin (or
  2 pouches liquid pectin)

1/2 cup fresh lemon juice
7 cups sugar

Stem elderberries and crush in small amounts in large pot. Bring to near boil, then simmer 15 minutes. To extract juice, place three layers of damp cheesecloth in strainer over bowl *or* place damp jelly bag in frame over bowl. Add cooked elderberries and allow to drip 4 hours or more, until dripping stops. Press gently on cooked berries to remove trapped juice. Measure out 3 cups juice.

*If using dry pectin,* add it to 3 cups juice, with lemon juice, in 8-quart pot. Stir and bring to boil. Add sugar and bring to full rolling boil. Boil 1 minute, stirring constantly.

*If using liquid pectin,* add sugar to juices, bring to boil, add pectin, bring to full rolling boil, and cook 1 minute, stirring constantly.

Both versions are alike from here on. Remove cooked jelly from heat, skim off foam, and pour into sterilized jars to within one-eighth inch of top. Wipe jar rims and threads. Cover quickly with lids and screw bands on tightly. Place in boiling water bath 5 minutes. Remove jars and allow to set.

### ELDERBERRY PIE

2 cups elderberries
2 tablespoons flour
1 cup powdered sugar

1 cup sour cream
graham cracker crust

Preheat oven to 350 degrees. Combine first four ingredients, mixing well. Pour into graham cracker crust. Bake 30 minutes.
*Jean M. Whiting, Cumberland, Maine*


### ELDERBERRY PIE

pastry for 2-crust pie (see
  juneberry recipes)
2¹/₂ cups stemmed elderberries
¹/₂ cup sugar

¹/₈ teaspoon salt
2 tablespoons flour
3 tablespoons lemon juice

Preheat oven to 450 degrees. Line pie pan with pastry. Fill with elderberries. Mix together sugar, salt, and flour; sprinkle over berries. Add lemon juice. Cover with top crust. Bake 10 minutes; reduce heat to 350 degrees and bake 30 minutes longer.
*Dot Heggie, Falmouth, Maine*


### ELDERBERRY WINE

2 gallons elderberries
6 pounds sugar

2 gallons boiling water
2 cakes yeast


Bruise fruit. Add sugar and boiling water and stir to dissolve sugar. Let set in crock 3 days. If fermentation has not started in that time, strain off juice and add yeast. Mix well. Let ferment 2 weeks, then strain and bottle.
*Phyllis Hobson,* Making Your Own Wine, Beer, & Soft Drinks: A Garden Guide Of Homestead Recipes

## SPICED ELDERBERRIES

1 stick cinnamon
1 tablespoon whole allspice
1 tablespoon whole cloves

1 pint diluted vinegar
3 pounds sugar
5 pounds elderberries

Tie spices in cheesecloth bag. Heat sugar, vinegar, and spices to boiling and cool. Add cleaned berries, heat slowly to simmering. Simmer until berries are tender. Cool quickly and let stand several hours or overnight. Remove spice bag. With slotted spoon separate berries from syrup. Pack berries in sterile jars, heat syrup to boiling, and pour over berries. Seal with two-piece metal lids and process in boiling water bath 15 minutes.

*Emily Krumm, Eaton Rapids, Michigan*

# Wild Grapes

The wild grape was one wild fruit my mother didn't harvest. We had domestic grapes, and they were a darned sight bigger than the wild varieties. Also, domestic grapes have more flesh than seeds; with wild grapes it's just the opposite.

When I found that wild grapes were relatively easy to pick and had such a full taste, I started harvesting them. I pick the entire clump and wait until I get home before pulling each grape off the stem.

My sons love wild grape jelly. When I visit them in the fall at the end of my fly-fishing guide work, they inevitably ask if there were many wild grapes. My response always varies—I'll answer them none, a few, or a bumper crop. If I answer bumper crop, they light up, knowing that I have put up some jelly for them.

Wild grapes are easy to pick if you find the right vines—for instance, those that are drooped over low-growing shrubs. If the vines are growing on a tall tree, you've got problems. One of the heaviest concentrations of wild grapes I've ever seen was growing on wild rose. That thorny rose made picking those grapes far worse than picking blackberries.

Wild grapes can be fairly easy to find. Your chances of finding enough for a batch of jelly are better than three in five. If you find a good grape patch, you'll have no trouble picking five to ten pounds.

Make sure to take along a pocket knife or a pair of shears to cut the grape clusters off the vine. Some people even take a ladder, since wild grape is a climber. My good friend Sam Ristich, an active forager, contends that wild grapes rate an 8.5 on a scale of 10 for climbing ability.

By the way, you don't need a bucket to pick wild grapes. You can put them in a plastic grocery bag (that's one way to recycle that darned plastic) or in a paper sack. I don't recommend putting them in your pocket—grape stains are forever. (You can remove grape stains with chlorine bleach, however.)

Wild grapes grow along almost all New England streams. Its vines climb the trees, other vegetation, and objects along the streams and in open woodlands. The leaves are deeply lobed, light green in color, and are about the size and shape of a typical sugar maple leaf. They are arranged alternately.

Wild grapes mature between late September to early October with ten to twenty pea-sized grapes per cluster. The grapes are deep blue to black with a blush. While they are very seedy, they make great jelly or jam. Their flavor is much stronger than commercial varieties, such as Concord.

Wild grapes are abundant in most of New England. Sometimes you will find grape vines entirely covering a tree or shrub. Grape vines have light brown bark, which is shredded into "strings." The vines have forked tendrils growing from the nodes (joints). The pith of grape stems is brown—an identifying feature.

Wild grape juice can be canned or frozen so that you can use it for jelly or other recipes at any time—and it's especially welcome in midwinter.

# Recipes

### VENISON JELLY

1 peck wild grapes
1 quart vinegar
1/4 cup cloves

1/4 cup stick cinnamon, broken
6 pounds granulated sugar

Put first four ingredients in preserving kettle. Heat to boiling and cook until grapes are soft. Strain through jelly bag. Boil juice 20 minutes. Add sugar and boil 5 minutes or until mixture jells. Pour into glasses and cover with paraffin (or process in boiling water bath per USDA guidelines).
*Helen Fitzgerald, Lansing, Michigan*

### WILD GRAPE JELLY

3 pounds wild grapes, crushed
1/2 cup water
1 cup apple juice

7 cups sugar
1 pouch liquid pectin

Mix crushed fruit and water. Bring to boil; cover and simmer 10 minutes, stirring occasionally. Extract juice using damp jelly bag. Then press gently. In 6- to 8-quart kettle, put 3 cups grape juice, sugar, and apple juice. Stir and bring to a full boil. Stir in pectin and bring to full rolling boil for 1 minute. Remove from heat, skim, fill jars to one-eighth-inch of top. Wipe rims and threads, attach lids, and screw on bands tightly. Put in boiling water bath for 5 minutes.
*Emily Krumm, Eaton Rapids, Michigan*

## WILD GRAPE JELLY (USING HONEY)

3¹/₂ cups honey
2¹/₄ cups grape juice
1 package pectin

¹/₄ teaspoon margarine or
butter

1. Measure honey in bowl. Set aside.
2. Measure grape juice into 6- or 8-quart saucepan or kettle. If a little short of juice, add water. If short more than 1 cup juice, add another type of fruit juice.
3. Add package of pectin to measured fruit juice. Stir thoroughly to dissolve, scraping sides of pan to make sure all the pectin dissolves. (This takes a few minutes.) Place mixture over high heat. Bring to a boil, stirring constantly to prevent scorching.
4. Add honey and mix well. Continue stirring and bring to full rolling boil (a boil that cannot be stirred down). Add margarine or butter and continue stirring. Boil hard exactly 2 minutes.
5. Remove mixture from heat. Skim foam and pour into glasses. Securely tighten two-piece metal lids, and process in boiling water bath 5 minutes.

Yield: five 8-ounce jars

MCP Foods, *Borden, Inc., Anaheim, California*

## SUPER-SMOOTH GRAPE JAM

3 to 4 quarts wild grapes to
   yield 4 cups pulp
1/2 cup water

3 cups sugar
1 lemon (juice and grated
   rind)

Wash and pick over grapes that are on verge of ripeness, including all green ones. Stem grapes. Place in a 6- or 8-quart saucepan or kettle. Add water and cook gently. Mash with wooden spoon until thoroughly softened. Strain mush through food mill to remove seeds. Measure 4 cups grape pulp and place in saucepan or kettle along with sugar, lemon juice, and grated rind. Cook gently, stirring to prevent burning, until mixture is thick enough to spread (about 20 minutes). Pour into hot sterilized jars, seal, and process 10 minutes in boiling water bath.

Yield: six 8-ounce jars

*Adapted from Clarissa M. Silitch, ed.* The Forgotten Arts: Making Old-Fashioned Jellies, Jams, Preserves, Conserves, Marmalades, Butters, Honeys & Leathers

# Juneberries

I find that picking juneberries can be a delightful way to spend a July day. Most of the times I've gone after juneberries, however, it has rained. One time that it didn't rain, I was with my buddy, Dick Appel.

We found a nice patch of juneberry. The bushes were six to fifteen feet tall. I could pick the smaller bushes quite easily, but when it came to the larger ones, we had to work together. Dick would grab a branch and slowly bend it down. When it was within reach, I'd hang onto it with one hand and pick with the other. Dick held out the berry bucket whenever my hand would get full. Perhaps this poses the question, how many hands do you need to pick juneberries? In this case, the answer was four.

Juneberries blossom from late April to early May at the same time the fiddlehead ferns emerge and the wake robins (red trilliums) blossom. They are among the first tree or shrub flowers of spring, so if you spot them, you should note the location and return in late July for the harvest.

Juneberries have five-petaled white flowers that hang in clusters like miniature Chinese lanterns. Usually, there are four to ten flowers in a cluster.

The leaf shape varies according to the species: Some are round, others oblong, others oval, and some are lance-shaped. Most of the species have leaves with slight serrations (small teeth). The leaves are deep green on top and pale green underneath.

Juneberry bushes range from four to twenty feet high. They have gray, smooth bark that has a texture like young beech bark, but with stripes.

The fruits, called pomes, are about the size of a large pea and have a five-pointed crown—just like blueberries and apples. As a juneberry ripens, it changes in color from light green to fuchsia to deep blue-black.

Eight to ten species of juneberries occur in New England, and the differences among the species are slight. They all belong to the genus *Amelanchier*. If you are observant, you might be able to distinguish four species. Other common names include shadbush and serviceberry. Look for juneberries on forested hillsides, in well-drained, gravelly soils, sandy plains, borders of fields, and in other clearings.

This berry was a favorite of many Indian tribes. It is one of the main components of pemmican, a dried meat concoction that Indians used as a winter food (see recipes).

Since juneberries are prone to insect infestations and rust, you will need to scout around to find a harvestable crop. Without scouting, your chances of finding enough berries to make jelly are less than fifty-fifty; with scouting, your odds increase to a 75 percent chance of getting two or more quarts. If you are fortunate to get a berry bucket full of this mildly sweet berry, you'll have the makings for wonderful jelly, jam, or pie.

One friend who doesn't like to take the trouble to extract juneberry juice in order to make jelly just stems the berries, heats them with a little water, then mashes them, and makes the pulp into jam. It's the seediest jam I've ever eaten, but tasty nonetheless. The seeds are big enough to stick in your teeth, but small enough to digest without difficulty.

You can freeze whole juneberries—just put them in plastic freezer bags and pop them into the freezer. They'll keep their flavor for six months or so. You can make jelly or pies with the frozen berries just as you would with fresh ones.

# Recipes

### PEMMICAN

This recipe is included as an historical note—to show how ingeniously the Native Americans preserved food for the winter.

Slice raw meat thinly and dry in the sun. Pound to a paste or grind finely. Mix with melted fat, dried juneberries, blueberries, and currants. (Dried fruit may be whole, ground, or chopped.) Form into cakes or encase in suitable slender bag or casing.

Pemmican will keep for a long time at room temperature.
*Dottie Litchfield, Brunswick, Maine*

### EVELYN'S BEST PIE CRUST
#### (2-crust recipe)

| | |
|---|---|
| 4 cups flour | 1/2 cup water |
| 1 tablespoon sugar | 1 egg, beaten |
| 1 teaspoon salt | 1 tablespoon vinegar |
| 1 3/4 cups shortening | |

Mix salt, sugar, and flour in a bowl. Cut in shortening. Mix together water, egg, and vinegar. Add and mix thoroughly into the flour mixture. Chill in refrigerator at least 15 minutes. Roll out on well-floured board using well-floured rolling pin.
Yield: 4 crusts (enough for two 2-crust pies)
*Evelyn Hejde, Aladdin, Wyoming*

## JUNEBERRY PIE

2 tablespoons flour
1/2 to 3/4 cup sugar
2 to 2 1/2 cups juneberries
pastry for 2-crust pie (see
above)

2 tablespoons lemon juice
1/4 cup water
2 tablespoons butter
4 tablespoons cream
2 tablespoons sugar

Preheat oven to 350 degrees. Mix together flour and sugar. Stir in juneberries. Line pie plate with pastry. Add berry mixture, then sprinkle with lemon juice and water. Dot with butter. Cover with top crust and prick in several spots. Brush with cream and sprinkle with sugar. Bake until juice boils up thickly.

*Irene Willey, Hulett, Wyoming*

## JUNEBERRY JELLY

1 pound juneberries
1 cup water
4 1/2 cups sugar

1/2 cup lemon juice
1 package pectin

Wash and crush ripe juneberries. Add water and simmer 15 minutes. To prepare juice: rest colander in bowl or kettle. Spread cloth or jelly bag over colander. Place hot prepared fruit into cloth or bag. Fold cloth to form bag and twist from top. Press with masher to extract juice.

Measure 3 cups juice into 6- to 8-quart kettle and add lemon juice and pectin. Stir well. Place over high heat, bring to boil, stirring constantly. Add sugar. Stir in well and continue stirring constantly. Bring to full rolling boil. Boil hard exactly 2 minutes. Remove from heat. Skim foam and pour into glasses. Seal with paraffin or use dome seals and boiling water bath method.

Yield: eight 6-ounce glasses

## JUNEBERRY MUFFINS

2 cups flour
1 teaspoon salt (scant)
1/4 cup sugar
3 teaspoons baking powder
1 egg, well-beaten

1 cup milk
1/3 cup vegetable oil
1 cup juneberries, fresh,
   frozen, or canned (if
   canned, drain well)

Preheat oven to 400 degrees. Combine dry ingredients. Stir and set aside.

In a large bowl beat egg. Add milk, oil, and berries. Stir gently to preserve whole berries. Add dry ingredients. Stir just to moisten. *Do not overmix.* Place in greased muffin tins and bake for 20 minutes.

Yield: 12 muffins

*Barb Griffith, Helena, Montana*

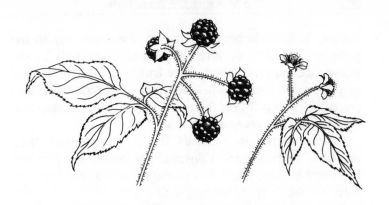

# Red Raspberries

New England has so many red raspberries, yet it appears that only the birds and animals enjoy them. It has always amazed me to see miles of red raspberries growing along the logging roads in Maine.

One day Dot and I discovered an old road lined on both sides with a thirty-foot band of ripe red raspberries for its entire three-mile length. We couldn't begin to pick that many berries, but we easily picked two quarts to take back to friends. While I was picking, I ate so many I thought I would burst. What a way to go. I can see the obituary headline now: "Man Bursts While Raspberry Picking—He Died with a Smile on His Face."

Wild raspberries deserve more attention from berry pickers than they get. They have a delicious taste and grow in profusion in many New England counties. Maybe berry pickers prefer the cultivated varieties of red raspberries due to the larger size of the cultivars, but berry pickers must remember that great taste comes in small packages!

Raspberries bloom from late May through mid-June. They mature rather quickly, ripening in early July and lasting through late July.

Ripe red raspberries look like smaller versions (i.e., red,

hollow, half-hemispheres) of the ones you can buy in the supermarket. (And definitely a lot less expensive!) To my knowledge, no other plants have fruits that resemble a red raspberry. Remember, when you pick raspberries the cores (receptacles) remain on the bush. When you pick blackberries, you get the receptacle and all.

Red raspberries are biennials, that is, they complete their life cycle in two years. Consequently, when you find red raspberry bushes, you'll find young canes without fruits on them as well as canes bearing fruit.

Red raspberries have a compound leaf composed of three or five leaflets. (When there are five leaflets, they are arranged pinnately, that is, the leaflets don't originate from the same point.) The canes may be prickly, bristly, or nearly smooth. The prickles may be straight or hooked. Some prickles may occur on the leaf stem.

Wild raspberry bushes range from two to five feet in height. The mature canes have a woody exterior and a pithy center.

Wild raspberries grow in well-drained soils. They prefer open areas. They often are one of the first inhabitants of a logged or burned site. Wild raspberries also grow along highway and railroad rights-of-way and on abandoned farms. As the vegetation matures into a forest, wild raspberries lose vitality and gradually die out.

Some of the best wild raspberry patches in New England grow around Grand Lake Stream, Maine. They flourish in recently cut timber stands and along logging roads. The ripe raspberries form fields of red during July. It's an easy matter to pick a couple of quarts of berries in an hour or so.

Most other areas in New England also have plenty of red raspberries. Most people would have few problems picking two quarts of raspberries in a two-hour period. Raspberries are affected by drought conditions, so if it has been dry, seek out raspberries growing on moister sites. Then you'll be picking the well-watered, juicy ones, not the dried up, drought-stricken ones.

Raspberries are easy to store. You can freeze them after washing and draining them or you can add ³/₄ cup sugar to each quart of raspberries or you can freeze them in a syrup. Either way, the raspberries will still have their great flavor when you pull them from the freezer in the middle of the winter.

# Recipes

Red raspberries are great snack fare. What a wonderful way to hike—along a trail lined by ripe, wild raspberries! You can eat your way to your destination. Wild raspberries are excellent with milk and sugar, on cold cereal, and as toppings for ice cream, pancakes, and waffles.

Wild red raspberries make the world's best jam with lots of seeds to stick in your teeth. A handful mixed into pancake batter makes for luscious pancakes.

Raspberries and blueberries combined make a wicked good pie.

### RED RASPBERRY JAM

2 quarts red raspberries
6¹/₂ cups sugar

1 pouch liquid pectin

Crush raspberries in small amounts. Measure out 4 cups crushed berries and place in 8-quart pot. Add sugar and bring to boil, stirring constantly. Add pectin and bring to rolling boil. Boil 1 minute, stirring constantly. Remove from heat and skim off foam. Pour into sterilized jars and wipe jar rims and threads. Cover quickly with lids and screw on bands tightly. Place in boiling water bath 5 minutes, then remove. Jars should seal within about 15 minutes. Jam can be stored indefinitely at room temperature after jars seal.
*Dot Heggie, Falmouth, Maine*

## RASPBERRY FOOL

1 pound fresh raspberries
3/4 cup sugar (or to taste)
1 cup heavy cream

2 tablespoons confectioners'
  sugar
3 tablespoons raspberry
  brandy

Cook berries and sugar over low heat until soft. While hot, press through sieve. When cooled completely, whip cream. Fold in sugar and brandy. Purée using a spatula. Spoon into small bowls or glasses. Chill. Serve *very cold*. Serve with butter cookies or pound cake.

Yield: 4 servings

*Janet Belanger, Buckfield, Maine*

## RASPBERRY RICOTTA PIE

pastry for 1 crust
2 1/2 cups fresh raspberries
3/4 cup sugar
1 1/2 tablespoons quick cooking
  tapioca
1/2 teaspoon cinnamon

8 ounces ricotta cheese
1 egg, separated
1/4 teaspoon salt
1/2 cup half and half
1 tablespoon lemon juice
3/4 teaspoon grated lemon peel

Preheat oven to 425 degrees. Line 9-inch pie plate with pastry. In bowl combine berries, sugar, tapioca, and cinnamon, until blended. Let stand 5 minutes.

Place in a blender: ricotta, egg yolk, salt, half and half, remaining sugar, and lemon juice plus peel. Blend until purée.

In a separate bowl beat egg white until soft peaks form. Fold into cheese mixture, just enough to blend.

Spoon berry mixture into shell, then spread cheese mixture over fruit. Sprinkle with cinnamon. Bake for 10 minutes, then reduce heat to 350 degrees for another 30 minutes, or until topping appears firm when you shake dish gently.

Yield: 6 servings

*Janet Belanger, Buckfield, Maine*

## RED BERRY MOUSSE

1 envelope unflavored gelatin
2 tablespoons cold water
juice and grated zest of 1 lemon
1 pint raspberries (save few for garnish)
1 pint strawberries, quartered

2 tablespoons syrup of Cassis
2 egg yolks
1/2 cup granulated sugar
2 cups heavy or whipping cream
1 bunch fresh mint

In small saucepan soak gelatin in water 5 minutes. Add lemon juice, stir, then add zest, raspberries, strawberries, and Cassis. Bring gently to a boil, stirring often. Remove from heat and cool to room temperature.

In small bowl beat egg yolks and sugar until pale yellow.

Place egg mixture in top of double boiler over simmering water. Whisk until slightly thickened and hot. Cool to room temperature. Fold into berry mixture until well blended.

Whip cream to soft peaks and gently fold into berry mixture, until well blended. Place in large glass bowl and chill till set.

Garnish with reserved raspberries and sprigs of mint.

Yield: 6 servings
*Helen Abbott, Moultonboro, New Hampshire*


## RASPBERRY PIE

2 tablespoons minute tapioca
1 cup sugar
1/4 teaspoon salt
4 cups fresh raspberries (or blackberries)

Pastry for 2 crusts
2 tablespoons butter or margarine

Preheat oven to 425 degrees. Combine tapioca, sugar, salt, and berries. Let stand 15 minutes. Line pie plate with bottom crust. Add berry mixture and dot with butter or margarine. Cover with top crust. Bake for 55 minutes, or until well browned. Serve warm or cold.

*Janet Belanger, Buckfield, Maine*

## CHILLED RASPBERRY SOUP

2 cups fresh raspberries
1/2 cup sugar
1/2 cup sour cream
2 cups ice water

1/2 cup red wine
4 or 5 small, fresh nasturtium leaves (optional), plus whole berries for garnish

Rub berries through fine sieve into bowl. Add sugar and sour cream and mix well. Add water and wine and mix well. Taste mixture to see if it needs additional sweeteners. Chill. (Let the mixture chill for several hours for full flavor.) Before serving, garnish with nasturtium leaves, if desired, and whole berries.
Yield: 4 to 5 servings
*Janet Belanger, Buckfield, Maine*

## CHERRY-RASPBERRY CONSERVE

3 cups tart red cherries
3 cups raspberries
4 1/2 cups sugar

1/2 cup chopped, blanched almonds or other nuts

Cook pitted cherries in very little water, about 1/3 cup, until tender. Add raspberries and sugar and cook until mixture is thick and clear. Stir in nuts and cook 5 more minutes. Pour into sterilized jars, wipe rims and threads clean, attach two-piece metal lids firmly, and place in boiling water bath 5 minutes.
*Adapted from Clarissa M. Silitch, ed.,* The Forgotten Arts: Making Old-Fashioned Jellies, Jams, Preserves, Conserves, Marmalades, Butters, Honeys & Leathers

## RASPBERRY-CURRANT MARMALADE*

8 cups raspberries                    9 cups sugar
4 cups currants

Wash and drain red or black raspberries before measuring. Stem and crush the currants. Cook slowly until juice flows freely. Add the raspberries and bring to boil. Add the sugar and boil hard to the jellying point.

(The jellying point is defined as 8 degrees above the boiling point of water where you live. If you have a candy thermometer, take the temperature of the water as it boils. The jellying point for most of New England would be 220 degrees Fahrenheit. For some of the mountainous areas, it will be higher. If you don't have a thermometer, you can use the spoon test. About 5 minutes after you add the sugar, take a metal spoonful of the jelly or jam and cool a minute. Holding the spoon a foot or so above the kettle, tip the spoon so the liquid runs back into the kettle. If it runs together at the edge and "sheets" off the spoon, the jelly is ready. Or put a saucer of the juice into your freezer. If the mixture firms up in a couple of minutes, the jelly is ready to pour into jars.)

Pour into sterilized jars, wipe jar rims and threads, attach two-piece metal lids securely, and place jars in a boiling water bath for 5 minutes.

*Adapted from Clarissa M. Silitch, ed.* The Forgotten Arts: Making Old-Fashioned Jellies, Jams, Preserves, Conserves, Marmalades, Butters, Honeys & Leathers

*You can use black raspberries or blackberries with this recipe.

## RASPBERRY SQUARES*

1 cup flour
1 teaspoon baking powder
1/2 cup butter
1 egg
1 tablespoon milk

2 cups shredded coconut
walnut-sized piece of butter
1 teaspoon vanilla
1 egg, beaten
1 cup sugar
raspberry jam

Preheat oven to 350 degrees. Sift together flour and baking powder. Cut in butter. Add egg and milk. Spread in greased 8 x 8 pan. Cover thinly with raspberry jam. In separate bowl, combine shredded coconut, butter pieces, vanilla, beaten egg, and sugar. Spread over jam. Bake for 30 minutes. Cut while warm.

Cooking to Beat the Band. *Compiled by Band Mothers Club, Deering High School, Portland, Maine*

*Can be made with black raspberries and blackberries, too.

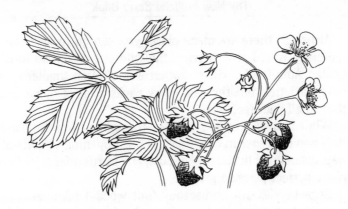

# Wild Strawberries

My first experience with wild strawberries came when my Dad, my brother, Jim, and I were on a trout-fishing trip. We had driven to a small stream, pitched our tent, and started a five-day vacation. After I had fished for two days from daylight to dark, I needed a break, so during midday I took a walk across a sand barren.

The area lacked trees and had few shrubs or grasses but it did have extensive patches of wild strawberries. While the strawberries were not much bigger than peas, they were plentiful and I managed to pick two or three cups.

The next morning, I proudly displayed the strawberries. Dad and Jim didn't say too much; they just poured them on their cold cereal, added a little sugar, poured on some milk, and started gobbling. The following morning we had strawberries sprinkled over our pancakes. Those strawberries made that fishing trip a lot more enjoyable. They added some variety to an otherwise monotonous menu of beans and hot dogs.

Wild strawberries still make welcome additions to a camper's menu. Strawberries are great for quick snacks while you're hiking, fishing, or strolling around camp. About the only preparation needed is to wash them because they do tend to pick up grit.

Although there are many places you can go to pick commercially raised strawberries, the wild ones seem so much tastier. Admittedly, wild strawberries are much smaller than tame varieties, but the wild berry packs a lot of flavor into that small package.

While strawberries are common throughout New England, they don't prosper in deep shade. Forest openings, rights-of-way, abandoned fields, timber cuts, and burns offer strawberries a perfect place to grow.

Strawberries are herbaceous (not woody) biennials that perpetuate the species by seeds and runners (stolons). The compound leaf is composed of three deeply serrated (toothed) leaflets.

Strawberries blossom in May and ripen in late June through mid-July. Each berry is about twice as large as a pea.

It's difficult to pick more than a quart of wild strawberries in a reasonable time, so most of its uses are limited to garnishes or embellishments. One of my favorite recipes consists of a half cup of wild strawberries mixed into pancake batter. Fry the pancakes and embellish them with a few especially red-ripe berries. Then drench the pancakes with a syrup made by simmering strawberries, water, and brown sugar. Who says camp meals have to be boring?

You can mix strawberries with other berries to get a little more mileage out of them. As the strawberries are tapering off, the blueberries, black and red raspberries, juneberries, and black currants are just starting. You can combine these berries and make a good berry stew. Combining berries to make cobblers, upside-down cakes, syrups, and jellies makes good sense and better taste!

You can freeze strawberries easily. Wash and stem them, drain well. Place them in plastic lock-type bags and freeze. Try putting premeasured amounts (say, one to two cups) into each bag so you won't have to bother measuring the thawed berries when you need them for a recipe.

# Recipes

### STRAWBERRY SYRUP

This is a handy recipe for the camp cook. It goes well with pancakes. Mash 1/2 to 1 cup of wild strawberries in a saucepan. Add 1/2 cup of water and 1/4 cup brown sugar. Bring to a boil, then simmer for 15 minutes.

### STRAWBERRY PANCAKES

Again, this is a simple camp recipe. Use a pancake mix that only requires that you add water. I usually make enough pancake batter for three people to eat three saucer-sized pancakes, which is a cup plus of mix. When you have mixed up the pancake batter, add approximately 1/4 cup of wild strawberries that have been washed and the caps removed. Cook the pancakes as usual—you might need just a little more cooking oil than normal to keep the pancakes from sticking. Serve with the above strawberry syrup and you'll have pancakes better than any you could make or buy back home.

### JELLY FLUFF FROSTING

1 cup strawberry jelly              1/8 teaspoon salt
2 egg whites

Combine all ingredients in top of double boiler. Beat with mixer over rapidly boiling water 7 minutes, or until mixture stands in peaks and is smooth and free of bubbles.
Yield: frosting for 2-layer, 9-inch cake
Cooking to Beat the Band. *Compiled by Band Mothers Club, Deering High School, Portland, Maine*

## STRAWBERRY JAM

2 quarts strawberries
7 cups sugar

1 pouch liquid pectin

Wash strawberries and remove caps. Crush in small amounts. Measure 4 cups crushed strawberries and place in 8-quart kettle along with sugar. Bring to full boil, stirring constantly. Add liquid pectin. Boil for 1 minute, stirring constantly. Remove from heat. Skim off foam. Pour into sterilized jars. Wipe jar rims and threads. Cover with two-piece lids. Screw bands tightly. Put in boiling water bath for 5 minutes. Remove. Let stand. The jars should seal within 30 minutes.
Yield: 7½ cups
*Certo Recipe, General Foods Corp., Kankakee, Illinois*


## FRUIT BATTER PUDDING*

2 cups fresh wild strawberries
½ cup sugar
½ cup shortening
1 cup sugar
1 egg, well-beaten

1 teaspoon vanilla
2 cups flour
2½ teaspoons baking powder
¼ teaspoon salt
1 cup milk

Preheat oven to 350 degrees. Mix berries with ½ cup sugar and put in well-greased square pan. Prepare batter: cream shortening well, and add 1 cup sugar. Add egg and vanilla. Beat until well blended. Sift dry ingredients together. Alternately add dry ingredients to batter with milk. Pour batter over berry mixture. Bake about 45 minutes.
*Emily Krumm, Eaton Rapids, Michigan*

*This recipe can be used with blueberries, cranberries, red and black raspberries, and blackberries as well.

## STRAWBERRY FRUIT ROLLS

4 cups strawberry purée      1 package pectin
(about 2 quarts whole     $^1/_2$ to 1 cup sugar
berries)

1. Use fully ripe strawberries. Wash and cut away any bruised or spoiled portions.
2. Purée strawberries in blender or food processor.
3. Stir the pectin into the purée. Mix well. Add sugar and stir until dissolved.
4. Spray cookie sheet or dehydrator shelf with vegetable oil. Spread 1 cup purée in border pattern (that is, along border of sheet or tray). Smooth purée with rubber spatula or tilt cookie sheet to evenly spread purée. Refrigerate unused purée.
5. *For conventional oven:* Set temperature control at lowest temperature or 150 degrees. Two cookie sheets may be placed in the oven at the same time. Rotate trays after $1^1/_2$ hours. Drying will take approximately $2^1/_2$ to 3 hours.

   *For dehydrator:* Set temperature control at 140 degrees and dry for 6 to 10 hours.

   *For sun drying:* Use same techniques as would be used for drying fruits.
6. Rolls are done when slightly sticky to the touch, but dry and pliable.
7. Remove rolls from tray while still warm and either roll each one in one piece or cut them into 4- or 6-inch squares. Roll in plastic wrap. Rolls may be stored up to 4 months without refrigeration. For longer storage, place in refrigerator up to 1 year or in freezer.

Yield: four 18" x 14" rolls

*MCP Foods, Inc., Anaheim, California*

## WILD STRAWBERRY JAM (USING HONEY)

4 cups strawberry pulp
  (approx. 2 quarts)
6 1/2 cups honey
1/4 cup lemon juice

1 package pectin
1/4 teaspoon margarine,
  butter, or cooking oil

1. Wash, stem, and crush strawberries thoroughly until reduced to pulp.
2. Measure honey in bowl and set aside.
3. Measure fruit pulp into 6- or 8-quart saucepan or kettle. If a little short of fruit, add water. Add lemon juice.
4. Add the package of pectin to fruit in kettle. Stir thoroughly to dissolve, scraping sides of pan to make sure all the pectin dissolves. (This takes a few minutes.) Place mixture over high heat. Bring to a boil, stirring constantly to prevent scorching.
5. Add the premeasured honey. Mix well. Bring to full rolling boil (a boil that can't be stirred down). Add the 1/4 teaspoon of margarine, butter, or cooking oil and continue stirring. Boil hard until the temperature reaches 221 degrees on a candy or deep-fat thermometer.
6. Remove mixture from heat. Skim foam and pour into glasses. Securely tighten two-piece metal lids and submerge glasses in boiling-water bath for 5 minutes.

Yield: ten 8-ounce jars

Make a Honey of a Jam with Recipes from MCP Food *(Division of Borden, Inc., Anaheim, California)*

## CRUNCHY FRUIT MUFFINS*

1¹/2 cups whole-wheat flour
2¹/2 teaspoons baking powder
¹/2 cup wheat germ or rolled
   oats
¹/2 cup chopped nuts or
   sunflower seeds (optional)

2 eggs, beaten
¹/4 cup lemon or plain yogurt
¹/4 cup honey
1 cup fresh strawberries

Preheat oven to 400 degrees. In large bowl combine flour, baking powder, oats or wheat germ, nuts or seeds. In small bowl combine eggs, yogurt, and honey. Mix lightly. Add wild strawberries or other berries or chopped fruit. Combine wet and dry ingredients and mix until just moistened. Batter may be lumpy. Grease 12 muffin cups, fill two-thirds full. Bake for 20 to 25 minutes until golden.

*Note:* If using juicy berries such as strawberries, blackberries, or raspberries, add last and mix in lightly before adding batter to muffin tins.

*Janet Belanger, Buckfield, Maine*

*Can be made with other berries or chopped fruit.

# Berry Miscellany

### TRAIL SNACK BERRIES

Blackberry
Black Raspberry
Blueberry/Huckleberry

Juneberry
Red Raspberry
Strawberry

### PIE BERRIES

Blackberry
Black Raspberry
Blueberry/Huckleberry
Chokeberry

Cranberry
Elderberry
Juneberry
Red Raspberry

### JAM AND JELLY BERRIES

Beach Plum
Blackberry
Black Currant
Black Raspberry
Blueberry/Huckleberry
Chokecherry

Cranberry
Elderberry
Juneberry
Red Raspberry
Strawberry
Wild Grape

### VERSATILE RECIPES

Black Raspberry Cobbler
Blackberry Frozen Yogurt
Blueberry Crisp
Crazy Crust Blueberry Pie
Crunchy Fruit Muffins
Fruit Batter Pudding

Hobo Cookies
Jelly Roll
Oatmeal Berry Muffins
Raspberry Currant Marmalade
Raspberry Squares
Yogurt Huckleberry Pie

# Glossary

*Alternate* refers to leaf arrangement. There is only one leaf per node on alternating sides of the stem.

*Biennial* the plant completes its life cycle in two years.

*Bloom* or *Blush* a whitish, powdery covering of the fruit, berry, leaf, or twig.

*Calyx* part of a flower which is beneath the petals. Sepals comprise the calyx.

*Cane* a pithy stem which is found in raspberry or elderberry.

*Cyme* a flat, or nearly flat, topped flower cluster.

*Inferior Ovary* the flower parts that arise from the top of the ovary.

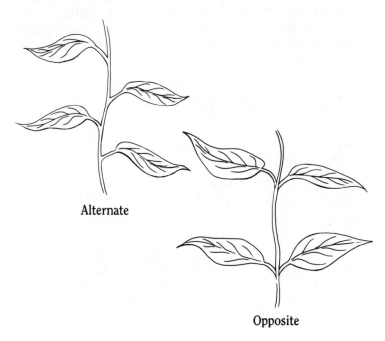

Alternate

Opposite

*Lenticels* breathing pores in the bark that resemble warts, or light-colored spots.

*Palmately Compound Leaf* the leaflets arise from a central point.

*Pinnately Compound Leaf* the leaflets arise along a central stem.

*Pome* fleshy fruit from an inferior ovary. Examples: juneberry and apple.

*Raceme* an inflorescence or cluster of flowers along 1 main stem.

*Sepal* part of a flower which is situated beneath the petals. Sepals comprise the calyx. Sepals are often green-colored.

*Serrate* having small teeth.

*Sheet Test* a cooking term that refers to the jellying point. Take a spoonful of hot jelly from the kettle, cool a minute, holding the spoon at least a foot above the kettle, tip the spoon so the jelly runs back into the kettle. If the liquid runs together at the edge and "sheets" off the spoon, the jelly is ready.

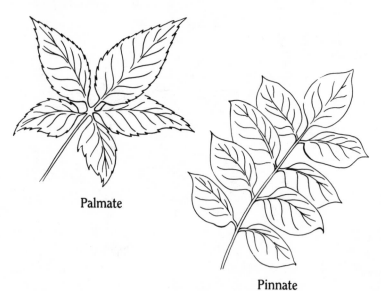

Palmate

Pinnate

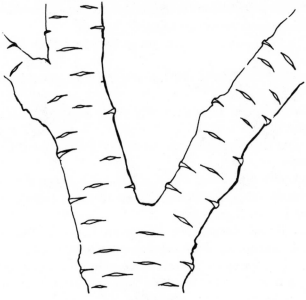

Lenticel

Umbel-like

# Bibliography

Certo Fruit Pectin, Homemade Jams & Jellies Directions. Kankakee, IL: General Foods Corporation.

Chesman, Andrea. *Summer in a Jar: Making Pickles, Jams & More*. Charlotte, VT: Williamson Publishing Co., 1985.

Cook, Mary Alice. *Traditional Portuguese Recipes from Provincetown*. Provincetown, MA: Shank Painter Publishing Co., 1983.

*Cooking to Beat the Band*. Portland, ME: Compiled by Band Mothers Club, Deering High School, 1940(?).

*Fresh Maine Blueberries Go Wild in Your Kitchen*. Augusta, ME: Maine Department of Agriculture, Food and Rural Resources.

Hobson, Phyllis. *Making Your Own Wine, Beer & Soft Drinks—A Garden Guide of Homestead Recipes*. Charlotte, VT: Garden Way Publishing, 1975.

*Huckleberry Recipes Compiled for Your Pleasure by the Swan Lake Women's Club*. Swan Lake, MT: Swan Lake Women's Club (no pub. date).

Hyland, Fay, and Ferdinand H. Steinmetz. *Trees and Other Woody Plants of Maine*. Orono, ME: University Press, 1944.

Lewis, Marion Averill. *The Norway Pines House Cookbook*. Old Town, ME: Penobscot Times, 1962(?).

*Make a Honey of a Jam With Recipes From MCP Foods.* Anaheim, CA: MCP Foods (no pub. date).

Pollard, Jean Ann. *The New Maine Cooking: Serving up the Good Life*. Augusta, ME: Lance Tapley, Publisher, 1987.

Richardson, Joan. *A Field Guide of New England Wild Edible Plants*. Chester, CT: The Globe Pequot Press, 1981.

Silitch, Clarissa M., ed. *The Forgotten Arts: Making Old-Fashioned Jellies, Jams, Preserves, Conserves, Marmalades, Butters, Honeys & Leathers*. Dublin, NH: Yankee Books, 1977.

*Special Recipes from MCP Foods.* Anaheim, CA: MCP Foods (no pub. date).

# Index

# About the Author

Born in farm country, Robert Krumm is a nature enthusiast who knows his berries. A professional fly-fishing guide, he has written for *Rod and Reel* and *Fly Fisherman,* and is nature editor for *Fly Fishing Heritage.*